International Perspectives on Aging

Volume 21

Series Editors

Jason L. Powell, Department of Social and Political Science, University of Chester, Chester, UK
Sheying Chen, Department of Public Administration, Pace University, New York, NY, USA

The study of aging is continuing to increase rapidly across multiple disciplines. This wide-ranging series on International Perspectives on Aging provides readers with much-needed comprehensive texts and critical perspectives on the latest research, policy, and practical developments. Both aging and globalization have become a reality of our times, yet a systematic effort of a global magnitude to address aging is yet to be seen. The series bridges the gaps in the literature and provides cutting-edge debate on new and traditional areas of comparative aging, all from an international perspective. More specifically, this book series on International Perspectives on Aging puts the spotlight on international and comparative studies of aging.

More information about this series at http://www.springer.com/series/8818

Irina Catrinel Crăciun

Positive Aging and Precarity

Theory, Policy, and Social Reality within a Comparative German Context

Springer

Irina Catrinel Crăciun
Department of Psychology
Babeş-Bolyai University
Cluj-Napoca, Romania

Department of Qualitative Social and
Educational Research
Freie Universität Berlin
Berlin, Germany

ISSN 2197-5841 ISSN 2197-585X (electronic)
International Perspectives on Aging
ISBN 978-3-030-14254-4 ISBN 978-3-030-14255-1 (eBook)
https://doi.org/10.1007/978-3-030-14255-1

Library of Congress Control Number: 2019933846

This Springer imprint is published by the registered company Springer Nature Switzerland AG
The registered company address is: Gewerbestrasse 11, 6330 Cham, Switzerland

Preface

The increase in longevity and the rise of aging populations have given rise to both concern and joy. Facing the challenge of investing longer lives with meaning and providing services for an increasing number of older individuals, researchers and policy makers have joined forces to develop solutions. Positive aging theories and active aging policies were formulated to design a framework for aging well. However, globalization, the spread of neoliberal ideas, labor market reforms, and the decline in social protection have altered the social context of aging and determined a rise in inequalities among older people (Grenier and Phillipson 2018; Grenier et al. 2017; Biggs 2014). The new culture of risk labeled as precarity (Grenier et al. 2017) makes positive aging less probable for certain social groups such as precarious workers. Several studies have dealt with the issue of precarity (Millar 2017; Standing 2011; Waite 2009; Butler 2004), and there is a growing evidence base on the links between precarity and aging (Grenier et al. 2017; Grenier and Phillipson 2018; Portacolone et al. 2018). The present book provides a novel psychological perspective on the connections between precarity and aging. Namely, it aims to understand what happens in the minds of precarious individuals and how people deal with existing discrepancies between positive aging ideals and the reality of precarious aging. The chapters are structured around questions such as "how does precarity shape aging experiences (i.e., views on precarity, views on aging, preparations for old age)?" and "is positive aging possible in precarious contexts?". The case studies that are used to illustrate the answers to these questions originate from a qualitative study conducted with precarious and financially secure middle-aged Germans. With its large aging population, adherence to active aging policies contrasted by rising numbers of precarious workers and inequalities among the elderly, Germany represents a fascinating case study. Nevertheless, since precarious aging represents a global phenomenon, findings provide scope for comparison with countries that are both similar and different from Germany in terms of welfare regimes and aging policies.

This book offers health psychologists, gerontologists, sociologists, social workers, health professionals as well as students and aging individuals themselves better insight into the meaning of precarious aging and builds confidence about aging well despite precarity. This work is dedicated to all precarious workers who find the courage to deal with life's uncertainties in a positive way.

Berlin, Germany Irina Catrinel Crăciun

References

Biggs, S. (2014). Precarious ageing versus the policy of indifference: International trends and the G20. *Australasian Journal on Ageing, 33*(4), 226–228.

Butler, J. (2004). *Precarious life: The powers of mourning and violence*. London: Verso.

Grenier, A., & Phillipson, C. (2018). Precarious aging: Insecurity and risk in late life. *Hastings Center Report, 48*, S15–S18. https://doi.org/10.1002/hast.907.

Grenier, A., Phillipson, C., Laliberte Rudman, D., Hatzifilalithis, S., Kobayashi, K., & Marier, P. (2017). Precarity in late life: Understanding new forms of risk and insecurity. *Journal of Aging Studies, 43*, 9–14. https://doi.org/10.1016/j.jaging.2017.08.002.

Millar, K. M. (2017). Toward a critical politics of precarity. *Sociology Compass, 11*(6), https://doi.org/10.1111/soc4.12483.

Portacolone, E., Rubinstein, R. L., Covinsky, K. E., Halpern, J., & Johnson, J. K. (2018). The precarity of older adults living alone with cognitive impairment. *The Gerontologist.* https://doi.org/10.1093/geront/gnx193.

Standing, G. (2011). *The precariat: The new dangerous class*. London: Bloomsbury Press.

Waite, L. (2009). A place and space for a critical geography of precarity? *Geography Compass, 3*(1), 412–433. https://doi.org/10.1111/j.1749-8198.2008.00184.x.

Acknowledgements

I am very grateful to the Alexander von Humboldt Foundation for awarding me a post doc research fellowship. This enabled me to conduct a research project that outlines the base of the present book. I am also thankful to Prof. Uwe Flick who was the host professor during my stay in Germany and provided feedback on the qualitative methodology of the study. My gratitude goes to all study participants. Without their commitment and motivation, this project would not have been possible.

A special thank you goes to Maria Crăciun for her timely and constructive critiques that have made an idea develop into a book. I am also very grateful to Silvia Făgărăşan and Hernan Biava for their valuable feedback and constant encouragement.

Last but not least, I would like to recognize the guidance and support of Prof. Chris Phillipson, the editors of the *International Perspectives on Aging Series*, Jason Powell and Sheying Chen, as well as the editors at Springer.

Contents

About the Author

Dr. Irina Catrinel Crăciun is Associate Professor in health psychology at the Babeş-Bolyai University in Cluj, Romania, and has been Guest Professor of health and gender at the Freie Universität Berlin. She also works as a health psychology practitioner and is Vice-President of the Romanian Health Psychology Association. She received her Ph.D. in health psychology at the Freie Universität Berlin (supervised by Prof. Dr. Ralf Schwarzer) and subsequently held a postdoctoral research grant on resources for positive aging, sponsored by the Alexander von Humboldt Foundation. Several articles resulted from this project and were published in journals such as *The Gerontologist*, *Journal of Aging Studies*, *Journal of Health Psychology*, and *Journal of Women and Aging*. Her research interests include a wide range of topics from health behavior change to health communication and exploring aging, gender, and psychosocial health determinants. Her recent studies focus mainly on aging and health perceptions in vulnerable populations such as aging precarious individuals, and the prevention of gender and social inequalities in health.

Chapter 1
Introduction: Issues of Aging and Precarity

Defining the Issue

Demographic changes are considered among the most relevant challenges of the twenty-first century. Public health achievements have led to an increase of around 30 years in life expectancy (Phillipson 2013). As a consequence, at a global level most industrialized countries can be described as aging societies, with rising percentages of older citizens all over the world (Wolff et al. 2018; Phillipson 2013). At an individual level, a longer life kindles the redefinition of old age to invest these later years with meaning. On a societal level, an increasing older population triggers a series of changes at cultural, institutional, and community levels. An aging population brings forward issues such as health insurance, pension plans, security, and housing for older persons. For example, new retirement policies have to be created to fit the growing number of potential retirees. EU policies encourage older people to work after their official retirement age, and strategies are devised to keep older people at work (Oude Mulders et al. 2016). In order to meet the needs of a rising older population and to foster policy change, a new way of thinking about old age was required. Against this background, the active aging policy emerged (Lassen and Moreira 2014; Walker 2009). The active aging policy goals include the following: to foster health in older people so that they do not become a burden for health systems, to keep older persons in employment, to reduce pension costs and to help seniors remain engaged in society (Stenner et al. 2011). From this perspective, aging is considered a lifelong process (Stowe and Cooney 2015), while "active" refers to continuing participation in activities across several domains from the economic, to culture and health (Walker 2006). Based on positive aging principles, the active aging policy encourages a healthy lifestyle (WHO 2002) and institutional reforms at EU level (Walker 2002, 2006, 2009; European Commission 1999, 2002, 2012). One way to realize both types of goals is by promoting novel representations of old age (Lassen and Moreira 2014; Stenner et al. 2011).

The active aging policy attempts to promote positive aging by changing views on aging both at individual level (i.e., perceptions of one's own aging) and at societal

I. C. Crăciun, *Positive Aging and Precarity*, International Perspectives on Aging 21, https://doi.org/10.1007/978-3-030-14255-1_1

level (i.e., perceptions of old people and old age). Positive views on aging were shown to influence health and longevity (Levy 2009; Levy et al. 2002) as well as engagement in health behavior (Wurm et al. 2010) and thus constitute a relevant resource for old age. In contrast, negative views on aging were found to be associated with ill health (Wurm et al. 2013) and disengagement in later life (Robertson and Kenny 2016). Therefore, having more positive views on aging could help better prepare for old age. However, positive views on aging may be more difficult to foster in a context of material deprivation and consequently, inequalities might occur in old age (Breheny and Stephens 2010).

All in all, with the decline of the welfare state and the widening inequalities among older individuals (Biggs 2014), new forms of insecurity arise (Grenier and Phillipson 2018; Grenier et al. 2017a). The concept of precarity represents a feeling of intrinsic uncertainty, resulting from coping with increasing pressures while trying to preserve a sense of independence (Portacolone et al. 2018; Grenier et al. 2017b; Portacolone 2013). It represents a universal form of insecurity, vulnerability, and potential suffering (Grenier et al. 2017a). The precariat was defined as a "class in the making" (Standing 2011) where different people can belong no matter their age, ethnicity, race, gender, or education level. Precarity implies new risks and responsibilities for aging individuals, specifically for the precariat. These novel risks (e.g., uncertain job contracts, temporary employment, no health insurance, etc.) and obligations (e.g., take responsibility for one's own pension and health) are reflected in the societal and individual views on aging (Craciun and Flick 2014; Craciun et al. 2015; Craciun and Flick 2016). Furthermore, views on aging depend on the social context where they emerge (Wolff et al. 2018). Apart from cultural differences (Fung 2013), social disparities can impact views on aging (Breheny and Stephens 2010). However, up till now, the research focus has been more on cultural variations (Löckenhoff et al. 2009) than on the social context related differences in views on aging. Exploring how precarity shapes views on aging can help understand how views on aging develop in a precarious social context as well as how to promote a positive old age by changing how people perceive it. Moreover, precarity may affect how people prepare for old age. By studying how precarity outlines the preparatory efforts, one can identify the obstacles that precarious individuals encounter when preparing for old age as well as the resources they may need in order to age well.

The positive aging policy with its emphasis on active engagement and planning may not match with the reality of aging in precarious times (Grenier et al. 2017a). Furthermore, previous research has shown that negative views on aging are prevalent in western societies (Levy 2009), and these negative perceptions of aging may become even more salient in precarious times (Craciun and Flick 2014). Thus, the evidence base presents mixed results concerning views on aging. For example, in Germany, there is an indication of an increase in positive views on aging (Beyer et al. 2017). However, these studies do not take into account the precarious context and how positive views on aging or preparations for old age may look like in precarious contexts. Moreover, one cannot but ask oneself how positive aging policy principles can become reality within the new culture of precarity (Grenier et al. 2017a).

Taking all the above into consideration, the present book aims to formulate some potential answers to questions such as "how does precarity shape the aging experience, views on aging and preparation for old age?" and "is there such a thing as positive aging in precarious times?"

Methods and Origins of the Book

The book is based on a research project on resources for positive aging, conducted by the author in 2012–2014 with the help of the Alexander von Humboldt Foundation. The aim of the study was to identify what resources for aging well are used by precarious individuals as compared to people with secure pension plans living in Germany. Berlin was chosen as a venue for the research since there are several aging precarious workers living in the capital (see Craciun and Flick 2014, 2016; Craciun et al. 2015). Germany itself as a country represented a good opportunity for a case study since it has a growing aging population (Mahne et al. 2017), a rising number of precarious jobs (Broding et al. 2010) and supports an active aging policy to promote positive views on aging (Tesch-Römer 2012).

Precarious work in Germany comprises temporary work, freelance activities, part-time jobs, the so-called mini jobs (i.e., paid around 450 euro a month) as well as temporary unemployment (Brady and Biegert 2017; Simonson et al. 2011). Among middle-aged Germans, women and those with low education or an immigration background are more likely to be part of the "working poor" (LaMontagne 2010; Bylow and Vaillant 2014). Nevertheless, the present study sample also included participants who are highly educated but are employed in precarious jobs. My aim was to obtain a variety of cases to allow for the exploration and illustration of the complex precarious reality.

Because views on aging are social cognitions, social representations, a good way to study them is by using episodic interviews (Flick 2014; Craciun and Flick 2014). This data collection method provides the opportunity to identify both semantic (e.g., knowledge about concepts or definition of relations between concepts) as well as episodic knowledge (based on concrete life situations) that make out the views on aging. For example, views on aging reflect an understanding of the concept of old age as well as of the process of aging. Episodic knowledge refers to the narratives, the contexts, and settings where this knowledge was acquired. Episodic knowledge is closer to experience because it refers to concrete events and situations when one thinks about aging or has experienced aging processes. For example, the setting of precarity provides participants with specific experiences of aging, that may be different from the positive aging scenarios disseminated by the active aging policy. Earlier studies (Flick et al. 2003; Flick et al. 2016) illustrated how the episodic interview is suitable for encouraging interviewees to talk about abstract definitions and beliefs and their influences, and also for contextualizing their experiences in real-life situations.

Concrete pointed questions were used in order to gather semantic information (i.e., *What does positive aging mean to you?*) or abstract relations between concepts (e.g., "*In your opinion what is the relation between positive aging and health?*"). Open-ended questions were formulated in order to elicit the narrative information or personal stories about aging (i.e., "Can you give me an example to illustrate what you associate with aging?"). Thus, in addition to asking participants to define positive aging (semantic knowledge), they were also asked to recount stories that would explain further what positive aging meant to them (episodic knowledge). This provided them with the opportunity to think about role models or situations when they were compelled to contemplate or prepare for positive aging. The choice of stories and their presentation (i.e., storytelling or argumentation) offered relevant information for the aims of this analysis. Also, each story that is told (episodic knowledge) is connected to a certain abstract concept (semantic knowledge). For instance, stories about old people who are positive role models are connected to general stereotypes of older people. All in all, the interview guide comprised eleven questions concerning people's perceptions of aging, representations of positive aging and resources for a good old age. The questions targeting representations of aging focused on people's subjective definitions of aging: "How can you tell that someone is old? Can you give me some examples?" and their perceptions of aging "What does aging mean to you? What do you associate with aging?" Other questions referred to contexts and situations when people think about aging "Is aging topic that you think about? If yes, can you give me some examples of situations when you think about aging?" as well as their concrete or desired preparations for old age.

For the present study, both participants with a secure pension plan (N = 20) and with an insecure pension plan (N = 20) were interviewed. The sample included 20 men and 20 women aged 45 to 65 years. Participants with a secure pension had a private or state pension plan or owned other financial assets in the form or real estate, business investments, etc. Precarious individuals were considered those who were freelancers, unemployed, or had a low-paid job and would have to rely on social security in their old age. None of the participants was retired at the time of the interviews. Participants were approached at their workplace (i.e., the participants who managed a shop or coffee place) or at leisure centers (i.e., dance schools, adult education centers, and community meeting centers) in Berlin neighborhoods which were chosen as representative for what used to be Western and Eastern parts of the city before the reunification of Germany. Other participants were recruited using snowball referrals (i.e., some participants recommended friends). Demographic data were gathered with a short questionnaire at the beginning of the interview. Interviews were conducted between November 2012 and August 2013 and lasted between 40 and 90 mins. Interviews were audio recorded and transcribed verbatim and reviewed for quality of the transcription.

The process of analysis was conducted following the principles of case-specific thematic coding (Flick 2014). This method was selected because it is recommended for comparative studies where the research question defines the groups that will be studied. The underlying assumption is that in different social groups, dissimilar views can be found regarding the research topic. First, all interviews were analyzed as a series of case studies. Each interview was described in terms of characteristics of the interviewed person, summary of what was said and a motto that represented the specific case. Starting from the analysis of one case, the thematic structure was formulated to include several dimensions that would serve to compare the groups. The comparison dimensions were as follows: definitions of aging, contexts of thinking about aging, representations about aging (i.e., fears, hopes, expectations, beliefs). Starting from these dimensions, all cases were analyzed and grouped together based on their main characteristics. All cases within one group were further analyzed based on their similarities. The analysis of cases between groups was focused on the differences existing between the groups (i.e., secure versus insecure pensions and men versus women). The resulting patterns were than analyzed and interpreted guided by the research questions.

A series of case studies were selected to explore the research questions formulated within the present book and illustrate the main arguments. These case studies will be presented as part of the argument in Chaps. 5, 6, and 7.

Summary of the Book

The ***introduction*** describes the issues that inspired the book, the main research questions, study goals, and methodology. Apart from the introduction, the book has three distinctive parts that are divided into three chapters each. Part I focuses on clarifying the research context drawing on theories of positive aging, the active aging policy recommendations, and social reality of precarious aging. Part II goes on to discuss the study results. The findings are organized in three chapters according to the research questions and explore what it means to be an aging precarious individual, what are the views on aging and preparations for old age in case of middle-aged precarious Germans. Part III examines the main discussions, conclusions, and implications for practice and policy of the present research.

Part I defines the main theoretical viewpoints and concepts that constituted the research background. It is structured around questions concerning the role of views on aging when formulating critiques of the successful aging model (Chap. 2), the link between active aging policy and views on aging (Chap. 3) and the connection between precarity and views on aging (Chap. 4).

Chapter 2 explores how positive aging theories shape views on aging. The chapter starts off with defining positive aging and goes on to provide an overview of the successful aging theory (Rowe and Kahn 1987; Kahn and Rowe 1998) and its main critiques. The key critical points are organized around the question of *whether we should think differently about aging*? Since theory inspires different views on aging, it is discussed how images of aging are modeled by theory and how they can be changed in accordance to critiques brought to the successful aging model.

The *first critique* addresses the issue of the successful aging model being outcome oriented. The point is made that images of aging, inspired by such a criteria-based model, are static and should be replaced by more process-oriented views on aging.

The latter could be informed by the selection, optimization, compensation model (SOC, Baltes and Baltes 1990), and the life course development approach (Joop and Smith 2006). If successful aging is defined as a management of resources (Freund 2008) across the lifespan, then views on aging can be regarded as a resource for positive aging that can be developed and used at different life stages.

A second critique concerns "success" as the core value of positive aging theories. Success is criticized for representing a capitalistic value, associating positive aging with a "busy ethic" that emphasizes activity and productivity in old age (Minkler and Holstein 2008). Images of successful old people can be used to make individuals the sole responsible for their old age and blame those who "fail" to age successfully in a youth obsessed culture (Rozanova 2010). Ageism and ableism are hidden behind such successful aging images of happy, healthy, active old people. Since successful aging criteria are difficult to achieve, many old people risk discrimination. From this critical viewpoint, cultural values (Torres 2006) provide different interpretations of success that are ignored by the mainstream successful aging theory. For example, the harmonious aging model is constructed as an alternative theory that takes cultural differences into account (Liang and Luo 2012). Starting from the premises of harmonious aging, culturally specific images of aging can be constructed as an alternative to success-oriented views on aging.

A third line of critique addresses the issue of successful aging being mainly theory driven and ignoring lay viewpoints of aging successfully. The "missing voices" perspective argues for taking individual definitions of positive aging into account. It is discussed how to use such lay perspectives of aging in order to promote novel images of aging that express individual variability. It is argued that various images of aging emerge not just in different cultures, but also diverse social contexts. One of the theses of the book is formulated, namely that a precarious social context may give rise to different positive and negative images of aging as compared to a social context of financial stability. The concluding section of the second chapter addresses the question of "*how to think about aging*?" The literature on positive and negative views on aging is reviewed, and potential changes that need to be made in thinking about aging are highlighted.

Chapter 3 examines the **politics of positive aging**. It starts off by defining active aging policy and exploring how positive aging theories helped shape aging policy. The chapter continues with an analysis of how active aging policy and images of aging are interlinked. It is argued that images of aging are not just theoretical constructs but play a significant role for aging policy development and implementation. Three examples are used to show how images of aging are used by policy: (1) the strategy used by the World Health Organization to support positive aging by encouraging behavior change at individual level, (2) the EU strategies applied to endorse institutional changes that can accommodate an increasing aging population, and (3) Germanies' active aging policy agenda that actively promotes positive aging images.

The fourth chapter discusses the **social reality of aging**, emphasizing the discrepancy between positive aging discourse and the precarious aging reality. The chapter starts out by depicting the characteristics of living in precarious times, referring to Baumann's (2007) ideas concerning "*living in liquid times.*" Precarity is defined as a characteristic of modern work conditions (Kalleberg 2009) but also a quality of human relations (Butler 2004; Portacolone 2013) and a global phenomenon of increased insecurity (Allison 2013). The implications of precarity for aging are discussed by examining the inconsistencies between the positive aging discourse and the social reality of aging. Several examples are used to illustrate this point: precarious work and "disappearing pensions," the insecurity associated with loneliness and health decline in old age and the "*precariatized mind*" as defined by Guy Standing (2011). The latter helps to demonstrate how precarity may shape cognitions and emotions.

Part II describes the research findings in an attempt to answer the main research questions: how precarious middle-aged individuals perceive the experience of precarity and aging (Chap. 5), what are their views on aging (Chap. 6) and how do they prepare for old age? (Chap. 7).

Chapter 5 discusses the precariat and outlines the characteristics of the German aging precariat, who represent the case study for the present book. A brief overview of the existing literature and statistics concerning the German precariat are provided to justify the relevance of the German case for explaining the link between precarity and aging.

Starting from the premise that precarity comprises both the vulnerability of aging bodies and minds as well as the intrinsic defenselessness of the shared social reality, Chap. 5 goes on and examines perceptions of precarity in middle-aged Germans. After summarizing the characteristics of the middle-aged individuals who participated in the study, the chapter focuses on the question of "*how one becomes precarious*?"

Results illustrate the paradox of freedom, namely how people have entered precarious jobs in search of freedom and ended up in the "precarity trap." Interviewees considered the age of 50 a "magic threshold" when people start thinking about precarity and aging. The chapter goes on to analyze the concept of the "precarious generation." Interviewees acknowledged they are aging and described themselves as a "surprised generation" who did not save for old age because they did not think they needed to do so. The concluding part of the chapter discusses how people cope with precarity. Several coping mechanisms for surviving precarity were mentioned by study participants. These coping strategies comprise the capacity to deal with risk (i.e., making courageous life choices) and the ability to find meaning in protest. From the participants' accounts a "precarious identity" emerges, characterized by a fighting spirit and flexibility. The link is made with the next chapter where views on aging are discussed by pointing out that the aging process provides new challenges concerning dealing with risks and opposing the insecurities of the precarious social reality.

Chapter 6 explores views on aging that emerge in a precarious reality. Since social reality is shared by individuals with precarious jobs and those who still enjoy financial security, the views on aging of the two groups are contrasted and compared. The chapter starts off by examining definitions of aging and situations where thoughts about aging occur. Old age was mainly defined in terms of decline and illness while thoughts about aging occurred in contexts where individuals encountered problems (e.g., parents' illnesses, personal health complaints, noticing aging signs in their friends, difficulties experienced at work because of older age, anxiety about the future). The body is constructed as an age identity card and a giveaway for a person's "true age." In contrast, how a person thinks and acts is considered to make one appear younger than one really is. Some participants go as far as to believe chronological age is obsolete, because beliefs and behaviors can make a person transcend their "real" age. Nevertheless, by concealing their true age, people also display a wish to hide how old they really are. The act of age concealment is important to individuals for distinct reasons. For precarious people, concealing their real age may protect them from losing their jobs and help them find work in old age. For financially secure individuals not showing the real age is important for avoiding the identification with the negative old age stereotype. Negative images of aging are predominant in both precarious and financially secure individuals, but they take different forms. For example, old age and aging are depicted by referring to symptoms and signs, very similar to an illness. The association of old age with illness and decline is characteristic for both precarious and financially secure study participants. Fears connected to aging are quite similar in the two groups and imply concerns related to dependency and illness. Nevertheless, while financially secure individuals fear loss of usefulness and attractiveness, precarious people fear they will not be able to work and sustain themselves in old age. Furthermore, avoiding thinking about aging (i.e., having no view on aging) is representative for precarious individuals.

The concluding section of the chapter deals with the possibility of positive images of aging within a precarious context. Financially secure people associate positive things with retirement, but not so much with aging itself. For them retirement implies a new-found freedom, having time for all the activities they postponed over the years. Nevertheless, precarious individuals also develop alternative positive images of aging. Despite their fears concerning old age, precarious people consider aging as a chance and regard wisdom as an important asset for old age.

Chapter 7 focuses on the actions that precarious and financially secure middle-aged Germans perform in preparation for old age. The first part of the chapter discusses the dilemma of precarious people who reflect whether preparatory efforts make sense in precarious times. However, precarious individuals do prepare for old age and their preparatory strategies include: thinking short-term, training their positive thinking capacity, investing in health, developing work-related skills that would help them get jobs in old age and developing social networks. Health, optimism, work, and family structures represent resources for a good old age in precarious times. The barriers toward reaching a positive old age are numerous and comprise a feeling of lacking control, the expenses related to a healthy lifestyle and lacking resources from the State. By contrast, financially secure individuals focus on planning as the main strategy for old age preparations. Among their preparatory strategies feature: training the body and mind to keep fit for later years, securing monetary assets to have financial independence in old age, leaving the big city and social activism to ensure meaning for old age. For financially secure persons, resources for old age include health, money, or real estate and even retirement itself because it provides a structure for their lives. In what concerns barriers for preparing old age, financially secure individuals do not really experience any which makes planning for old age an uncomplicated matter.

Part III discusses how the studies' main findings add to or challenge the existing literature and outline implications for practice and policy. The results discussion revolves around several questions: "*how does precarity shape views on aging?*" (Chap. 8), "*how does precarity shape preparations for old age*?" (Chap. 9) and whether "*positive aging can exist in a context of precarity*?" (Chap. 10).

Chapter 8 argues that precarity shapes views on aging by generating and encouraging specific negative images concerning old age, by stimulating the reinvention of positive aging images and by increasing the discrepancy between positive aging ideals and the social reality of precarity. First, the chapter examines the negative images forged by precarity, namely the different fears concerning old age that constitute an expression of precarity. Negative images of aging are linked to the precarity of the body such as old age represented as illness and the body constructed as an old age identity card. Several precarity related developments are further analyzed, such as the disappearance of old age, the youth culture and the desired agelessness, chronological age becoming obsolete and the denial of aging.

The chapter goes on to discuss how positive aging views are reinvented in a precarious context. The question of whether precarious middle-aged Germans hold positive images of aging is addressed. It is argued that precarious individuals create alternative positive images of aging that cannot be integrated in the mainstream active aging policy discourse.

Last, but not least, the chapter analyses the discrepancies between positive aging discourses and the precarious reality of aging as well as the consequences of these discrepancies. The point is made that negative images of aging are similar in precarious and financially secure individuals since both groups share the same precarious reality. The idea of clear social inequalities is criticized. Since people share the same precarious reality, they also have similar precarity-related problems even though financially secure individuals might feel protected by their material assets. The point is made that the "precariatized mind" is not just a characteristic of precarious individuals. For example, fears of aging are similar in precarious and financially secure individuals. However, their hopes and coping mechanism differ depending on the resources they have. In this respect, precarious individuals experience more barriers toward reaching a positive old age and could be considered at a disadvantage in what successful aging is concerned.

Chapter 9 considers the issue of preparation for old age in precarious times. It is argued that precarity shapes old age preparations by constructing the aging context, by affecting mental preparations and creating distinctive resources for old age.

First, the precarity context implies certain work and health conditions that change the way people prepare for old age. The link between negative images of aging fostered by precarity and lack of health preparations is discussed. For example, precarity can determine ageism in the work setting. Disadvantages accumulate and can determine inequalities in old age making precarity responsible for emerging social inequalities.

Second, mental preparations for old age are affected by the features of the precarious mind such as short-term thinking, the feeling of lack of control over resources, and the inability to plan. Precarity nurtures lack of agency and diminished confidence and threatens the self-identity concerning aging.

A third way in which precarity affects aging preparations is by encouraging the creation of novel resources for old age and finding alternative preparatory strategies. Preparation for old age is reinvented in the context of precarity. If the strategies recommended by active aging policy do not work in a precarious context, people must create their own ways to prepare for old age. Even if precarious individuals have mainly negative images of aging, they do prepare for old age. However, they do not choose the classical preparatory actions, dictated by mainstream positive aging discourses. For example, planning does not necessarily constitute an adaptive strategy in the precarious context. As an alternative, disengagement and relaxation may be useful strategies for precarious individuals. Discrepancies between preparatory actions proposed by successful aging models and the precarious reality are discussed to identify those strategies that can build an accord between theory, policy, and social reality.

Chapter 10 comprises the research conclusions and implications for practice and policy. The chapter argues that *positive aging can exist in precarious contexts*, because precarity shapes how middle-aged individuals imagine old age and how they enact preparations for old age. The research conclusions are framed by this discussion of how individuals manage to age happily within the new culture of risk. However, the fact that people instinctively find "ways to survive" precarity does not mean that they do not need help. On the contrary, the point is made that middle-aged precarious individuals require assistance both at individual and policy level to ensure their health and well-being in old age. Implications for practice and policy in relation to positive aging and health are outlined. For example, at individual level, counseling to avoid the "mind traps" of precarity as well as develop flexibility, confidence, and foster positive thinking patterns concerning old age may be effective. At policy level, interventions attempting to manage age discrimination on the job market, develop a "health for all" policy, and promote novel images of aging in the mass media are further considered in the chapter. Last, but not least, future research directions concerning the upcoming precarious aging generation are outlined, and the theoretical integration of precarity and precarious aging in the successful aging paradigm is discussed.

References

Allison, A. (2013). *Precarious Japan*. Duke University Press. https://doi.org/10.1215/9780822377245.

Baltes, P. B., & Baltes, M. M. (Eds.). (1990). *Successful aging*. Cambridge University Press. https://doi.org/10.1017/cbo9780511665684.

Baumann, Z. (2007). *Liquid times: Living in an Age of Uncertainty*. Cambridge: Polity Press.

Beyer, A.-K., Wurm, S., & Wolff, J. K. (2017). Älter werden – Gewinn oder Verlust? Individuelle Altersbilder und Altersdiskriminierung [Aging-Gain or Loss? Individual views on aging and ageism]. In Mahne et al. (Eds.), *Altern Im Wandel*, 329–343. https://doi.org/10.1007/978-3-658-12502-8_22.

Biggs, S. (2014). Precarious ageing versus the policy of indifference: International trends and the G20. *Australasian Journal on Ageing, 33*(4), 226–228.

Brady, D., & Biegert, T. (2017). The rise of precarious employment in Germany. *Research in the Sociology of Work*, 245–271. https://doi.org/10.1108/s0277-283320170000031008.

Breheny, M., & Stephens, C. (2010). Ageing in a material world. *New Zealand Journal of Psychology, 39*(2), 41–48.

Broding, H. C., Weber, A., Glatz, A., & Bünger, J. (2010). Working poor in Germany: Dimensions of the problem and repercussions for the health-care system. *Journal of Public Health Policy, 31*(3), 298–311. https://doi.org/10.1057/jphp.2010.20.

Butler, J. (2004). *Precarious life: The powers of mourning and violence*. London: Verso.

Bylow, C. and Vaillant, K. (2014). *Die verratene Generation Was wir den Frauen in der Lebensmitte zumuten* [The betrayed generation. What should we expect from middle-aged women?]. München: Pattloch.

Craciun, C., & Flick, U. (2014). "I will never be the granny with rosy cheeks": Perceptions of aging in precarious and financially secure middle-aged Germans. *Journal of Aging Studies, 29,* 78–87. https://doi.org/10.1016/j.jaging.2014.01.003.

Craciun, C., Gellert, P., & Flick, U. (2015). Aging in precarious circumstances: Do positive views on aging make a difference? *The Gerontologist*, gnv135. https://doi.org/10.1093/geront/gnv135.

Craciun, C., & Flick, U. (2016). Aging in precarious times: Exploring the role of gender in shaping views on aging. *Journal of Women and Aging, 28*(6), 530–539. https://doi.org/10.1080/08952841.2016.1223896.

Commission, European. (1999). *Towards a Europe for all ages*. Brussels: European Commission.

Flick, U. (2014). *Handbook of qualitative data analysis*. The Sage. https://doi.org/10.4135/9781446282243.

Flick, U., Fischer, C., Neuber, A., Schwartz, F. W., & Walter, U. (2003). Health in the context of growing old: Social representations of health. *Journal of Health Psychology, 8*(5), 539–556. https://doi.org/10.1177/1359105303008500 6.

Flick, U., Hans, B., Hirseland, A., Rasche, S., & Röhnsch, G. (2016). Migration, unemployment, and lifeworld. *Qualitative Inquiry, 23*(1), 77–88. https://doi.org/10.1177/1077800416655828.

Freund, A. M. (2008). Successful aging as management of resources: The role of selection, optimization, and compensation. *Research in Human Development, 5*(2), 94–106. https://doi.org/10.1080/15427600802034827.

Fung, H. H. (2013). Aging in culture. *The Gerontologist, 53*(3), 369–377. https://doi.org/10.1093/geront/gnt024.

Grenier, A., & Phillipson, C. (2018). "Precarious Aging: Insecurity and Risk in Late Life," *What Makes a Good Life in Late Life? Citizenship and Justice in Aging Societies,* special report, *Hastings Center Report* 48, no. 5, S15–S18. https://doi.org/10.1002/hast.907.

Grenier, A., Phillipson, C., Laliberte Rudman, D., Hatzifilalithis, S., Kobayashi, K., & Marier, P. (2017a). Precarity in late life: Understanding new forms of risk and insecurity. *Journal of Aging Studies, 43,* 9–14. https://doi.org/10.1016/j.jaging.2017.08.002.

Grenier, A., Lloyd, L., & Phillipson, C. (2017b). Precarity in late life: Rethinking dementia as a "frailed" old age. *Ageing, Dementia and the Social Mind*, 142–154. https://doi.org/10.1002/9781119397984.ch11.

Joop, D., & Smith, J. (2006). Resources and life-management strategies as determinants of successful aging: On the protective effect of selection, optimization, and compensation. *Psychology and Aging, 21,* 253–265. https://doi.org/10.1037/0882-7974.21.2.253.

Kahn, R. L., & Rowe, J. W. (1998). Dr. Rowe and Dr. Kahn reply. *The Gerontologist, 38*(2), 151. https://doi.org/10.1093/geront/38.2.151a.

Kalleberg, A. L. (2009). Precarious work, insecure workers: Employment relations in transition. *American Sociological Review, 74*(1), 1–22. https://doi.org/10.1177/000312240907400101.

LaMontagne, A. D. (2010). Commentary: Precarious employment: Adding a health inequalities perspective. *Journal of Public Health Policy, 31*(3), 312–317. https://doi.org/10.1057/jphp.2010.25.

Lassen, A. J., & Moreira, T. (2014). Unmaking old age: Political and cognitive formats of active ageing. *Journal of Aging Studies, 30,* 33–46. https://doi.org/10.1016/j.jaging.2014.03.004.

Levy, B. (2009). Stereotype embodiment: A psychosocial approach to aging. *Current Directions in Psychological Science, 18,* 332–336. https://doi.org/10.1111/j.1467-8721.2009.01662.x.

Levy, B. R., Slade, M. D., Kunkel, S. R., & Kasl, S. V. (2002). Longevity increased by positive self-perceptions of aging. *Journal of Personality and Social Psychology, 83*(2), 261–270.

Liang, J., & Luo, B. (2012). Toward a discourse shift in social gerontology: From successful aging to harmonious aging. *Journal of Aging Studies, 26*(3), 327–334.

Löckenhoff, C. E., De Fruyt, F., Terracciano, A., McCrae, R. R., De Bolle, M., Costa, P. T. Jr., … Yik, M. (2009). Perceptions of aging across 26 cultures and their culture-level associates. *Psychology and Aging, 24,* 941–954.

Oude Mulders, J., Henkens, K., & Schippers, J. (2016). European top managers' age-related workplace norms and their organizations' recruitment and retention practices regarding older workers. *The Gerontologist*, gnw076. https://doi.org/10.1093/geront/gnw076.

Mahne, K., Wolff, J. K., Simonson, J., & Tesch-Römer, C. (2017). Altern im Wandel: Zwei Jahrzehnte Deutscher Alterssurvey [Changes in aging: Two decades German aging survey]. In Mahne et al. (Eds.), *Altern Im Wandel*, 11–28. https://doi.org/10.1007/978-3-658-12502-8_1.

Minkler, M., & Holstein, M. B. (2008). From civil rights to … civic engagement? Concerns of two older critical gerontologists about a "new social movement" and what it portends. *Journal of Aging Studies, 22*(2), 196–204. https://doi.org/10.1016/j.jaging.2007.12.003.

Phillipson, C. (2013). *Ageing.* Cambridge: Polity Press. ISBN 13: 978 0 7456 3085 4.

Portacolone, E., Rubinstein, R. L., Covinsky, K. E., Halpern, J., & Johnson, J. K. (2018). The precarity of older adults living alone with cognitive impairment. *The Gerontologist.* https://doi.org/10.1093/geront/gnx193.

Portacolone, E. (2013). The notion of precariousness among older adults living alone in the U.S. *Journal of Aging Studies, 27*(2), 166–174. https://doi.org/10.1016/j.jaging.2013.01.001.

Robertson, D. A., & Kenny, R. A. (2016). "I'm too old for that"—The association between negative perceptions of aging and disengagement in later life. *Personality and Individual Differences, 100,* 114–119.

Rowe, J., & Kahn, R. (1987). Human aging: Usual and successful. *Science, 237*(4811), 143–149. https://doi.org/10.1126/science.3299702.

Rozanova, J. (2010). Discourse of successful aging in The Globe and Mail: Insights from critical gerontology. *Journal of Aging Studies, 24*(4), 213–222. https://doi.org/10.1016/j.jaging.2010.05.001.

Simonson, J., Gordo, L. R., & Titova, N. (2011). Changing employment patterns of women in Germany: How do baby boomers differ from older cohorts? A comparison using sequence analysis. *Advances in Life Course Research, 16*(2), 65–82. https://doi.org/10.1016/j.alcr.2011.03.002.

Standing, G. (2011). *The precariat: The new dangerous class*. London: Bloomsbury Academic.

Stenner, P., McFarquhar, T., & Bowling, A. (2011). Older people and "active ageing": Subjective aspects of ageing actively. *Journal of Health Psychology, 16*(3), 467–477. https://doi.org/10.1177/1359105310384298.

Stowe, J. D., & Cooney, T. M. (2015). Examining Rowe and Kahn's concept of successful aging: Importance of taking a life course perspective. *The Gerontologist, 55*(1), 43–50. https://doi.org/10.1093/geront/gnu055.

Tesch-Römer, C. (2012). *Active aging and quality of life in old age*. New York and Geneva: United Nations.

Torres, S. (2006). Different ways of understanding the construct of successful aging: Iranian immigrants speak about what aging well means to them. *Journal of Cross-Cultural Gerontology, 21*(1–2), 1–23. https://doi.org/10.1007/s10823-006-9017-z.

Walker, A. (2009). Commentary: The emergence and application of active aging in Europe. *Journal of Aging and Social Policy, 21*(1), 75–93. https://doi.org/10.1080/08959420802529986.

Walker, A. (2006). Aging and politics: An international perspective. In R. Binstock & L. George (Eds.), *Handbook of aging and the social sciences* (pp. 339–359). Burlington, MA: Elsevier Academic Press.

Walker, A. (2002). A strategy for active ageing. *International Social Security Review, 55*(1), 121–140.

Wolff, J. K., Wiest, M., Nowossadeck, S., Beyer, A., & Wurm, S. (2018). Regional impact of demographic change on changes in individual views on aging. *Innovation in Aging*, 1(suppl_1), 1335. https://doi.org/10.1093/geroni/igx004.4895.

World Health Organization. (2002). *Active ageing: A policy framework.* Geneva: World Health Organization.

Wurm, S., Tomasik, M. J., & Tesch-Römer, C. (2010). On the importance of a positive view on ageing for physical exercise among middle-aged and older adults: Cross-sectional and longitudinal findings. *Psychology and Health, 25*(1), 25–42. https://doi.org/10.1080/08870440802311314.

Wurm, S., Warner, L. M., Ziegelmann, J. P., Wolff, J. K., & Schütz, B. (2013). How do negative self-perceptions of aging become a self-fulfilling prophecy? *Psychology and Aging, 28*(4), 1088–1097. https://doi.org/10.1037/a0032845.

Part I

Chapter 2
Positive Aging Theories and Views on Aging

Introduction

As the candles gather on the birthday cake with the passing years, some people choose not to celebrate their birthdays anymore or even to change the "real" number of candles that should be on the cake. However, others do just the opposite and mark the occasion when they turn 80 or 90 as a big and joyful event. Why do people behave this way and how is it that such a potentially happy event as a birthday would not be celebrated anymore? Behind such differences one can find diverse expectations, fears, and hopes people hold concerning old age. Children or teenagers cannot wait to be older, and sometimes even pretend to be so. However, at some point in a person's life, it becomes relevant to be younger or at least appear younger than one's real, chronological age. Birthdays constitute a reminder of one's aging process, and the identity card becomes undisputable proof of one's real chronological age. With the adding years, happiness and anticipation concerning ones' birthday may turn either into melancholy and dread or into a feeling of satisfaction and accomplishment. But why does this happen and how can one influence it?

Birthdays have different meanings depending on what people think about aging and old age. The act of celebrating birthdays is actually a modern practice, since before the twentieth century people were not so interested in birthdays as representing rites of passage (Phillipson 2013). Birthdays have implications such as being "too old" for something, or "just the right age." These evaluations are influenced by how old age is perceived in general or how certain life stages are appraised (e.g., "being in your 20 s is wonderful" or "being 30 is better than being 20" or "being in your fifties is terrible"). Individual beliefs about old age, old people, and the aging process guide how persons see their own age or the aging process itself.

Aging and old age have captured the imagination of people over centuries, as numerous art pieces and a vast literature depicting older adults' attest. Fears and hopes about growing older are expressed in sayings such as "*one wants to be older but not old*" or "*one is only as old as one feels.*" Apart from a concern for medical issues related to old age, from a psychological perspective, there is a growing interest

I. C. Crăciun, *Positive Aging and Precarity*, International Perspectives on Aging 21, https://doi.org/10.1007/978-3-030-14255-1_2

for subjective aging, namely how people perceive their own aging or imagine their own old age (Westerhof et al. 2014).

Subjective aging includes several constructs such as self-perceptions of aging (what people think about their own aging), stereotypes about aging (beliefs about typical characteristics of old people and old age), age identity (identification with a certain age group), awareness of age-related changes and subjective age (how old a person perceives himself/herself to be). Among these constructs that can be included under the paradigm of subjective aging, views on aging play a key role since they embody self-perceptions of aging as well as societal stereotypes of old age and the elderly. Beliefs about aging have been conceptualized, measured, and labeled in various ways: self-perceptions of aging (Levy et al. 2002), expectations regarding aging (Sarkisian et al. 2002), aging stereotypes (Levy 2009), views on aging (Wolff et al. 2016), and attitudes toward older people. For the sake of coherence and simplicity, I will use a generic term calling them **"views on aging"** throughout this book.

Theory helps explain *how* and *why* views on aging are relevant for our aging process and *affect how* old persons are treated within a society. Individual views on aging are influenced by societal views on aging or aging stereotypes existing in a certain society or culture. Nevertheless, if one wants to delve deeper and ask what influences societal views on aging and old age stereotypes, one would discover the theoretical thoughts behind aging representations and stereotypes. According to the *stereotype embodiment theory* (Levy 2009), aging stereotypes are embodied when they are absorbed from the surrounding culture and become self-definitions of aging. Stereotypes about aging are internalized across the life span, can operate at an unconscious level, gain salience from self-relevance, and use multiple pathways to influence health and well-being (Levy 2009). Concerning the latter, I will discuss later in the present chapter how self-definitions or self-perceptions of aging influence the health and functioning of individuals. In the following, I will explore how positive aging theory shapes our views on aging, then go on to discuss why the way we think about aging is so important, as well as how views on aging may change along with theory development.

How Positive Aging Theory Shapes Views on Aging

Positive Aging—A Shift in the Research Paradigm

For many decades, old age had been regarded as a period of life, characterized by physical and psychological decline, and social withdrawal. In accordance with such definitions, the first conceptual models considered decline and dependence as inherent to old age. For example, the *disengagement theory* stipulated that as skills deteriorate with age and the remaining life is perceived to be shorter, people withdraw from their social roles and society engages them less (Cumming and Henry 1961).

The activity and continuity theories were the first to challenge the assumption that decline constitutes a normal aspect of aging. First, the *activity theory* (Havighurst 1961, 1963) postulated that staying active and socially engaged ensures life satisfaction in old age. Then, the *continuity theory* (Atchley 1999) stated that older adults seek to maintain their activities, roles, and relations because it gives them a sense of permanence and well-being. Nevertheless, neither the disengagement theories nor the activity and continuity theories could explain the heterogeneity encountered among aging individuals. Thus, in an attempt to explain why some people, age well while others do not, Havighurst (1963) proposed that satisfaction with ones' past and present life should be definitory for successful aging.

Early definitions of human aging with their emphasis on decline made it difficult to distinguish aging processes from disability or chronic illness (Dovenmuehle et al. 1970). Critiques pointed out that aging is not an illness, even though many old people suffer from one ailment or another. Palmore (1970) introduced the term "*normal aging*" to distinguish natural aging process from illnesses that may occur in old age because of stress, accidents, or maladjustments.

Large population-based studies like the Baltimore Longitudinal Study, that followed adults' development across the life span, showed that some people reached old age without major impairments to their health (Hill 2011). Furthermore, evidence from the Berlin Aging Study indicated that even when facing age-related decline, many people still experience well-being in old age (Kunzman et al. 2000). The discrepancy between the objective decline associated with old age and the maintenance or even improvement of subjective well-being was labeled "the paradox of aging" and inspired extensive research (Baltes and Baltes 1990, 1991; Staudinger et al. 1993; Carstensen et al. 1999; Löckenhoff and Carstensen 2004). Further studies pointed out that old individuals tended to rate themselves as aging well and considered it unlikely to face decline and loneliness in old age (Bowling 2005, 2007; Bowling and Dieppe 2005). A new theoretical approach was needed to explain the heterogeneity of aging experiences and the occurrence of subjective well-being in old age.

The emphasis in research shifted to studying the variability in old age and the ability of individuals to adapt to changes. Encouraged by this shift in thinking about aging, gerontologists asked themselves whether there is potential for development in old age. The pathological model of old age ignored the heterogeneity of the aging experience and could not explain subjective well-being in late adulthood. Thus, researchers looked for inspiration in positive psychology theories in order to develop the concept of positive aging. According to positive psychology, life transitions are interpreted to maximize well-being (Hill 2011). Nevertheless, in case of old age, many life transitions may be marked by age-related decline. For example, people must cope with unavoidable losses to maintain well-being at advanced stages in their life. In order to experience a positive old age despite illness, a person needs to remain optimistic even when facing physical and mental decline, loss, pain, or aging-related grief. In line with this assumption, Hill (2005) described positive aging by naming several characteristics: (1) the ability to mobilize inactive coping abilities, (2) flexibility of thought and behavior, (3) decision-making ability that affirms well-being even when one must give up pleasant activities that are not possible anymore, and (4) keeping an optimistic viewpoint regarding decline-related matters. In parallel with psychological studies, biomedical research also gradually started to recognize the distinction between aging processes and the progression of illness (Rowe and Kahn 1987; Vaillant and Mukamal 2001). All in all, health and well-being began to be regarded as "normal aging," while activity and social engagement were considered the key ways to achieving a good old age.

The Successful Aging Theory

Developed by Rowe and Kahn (1987, 1997), the successful aging theory rapidly grew in popularity because it made the possibility of a flourishing old age seem more real. This theory was also the first to introduce a clear distinction between "usual" and "successful" aging. The theory stipulates a series of criteria that a person needs to fulfill to be considered to have reached old age successfully. These standards include the following: avoiding disease and disability in old age, maintaining high cognitive and physical functioning, and keeping an active engagement with life (Rowe and Kahn 1997). All components of the model are relevant for deciding whether a person is aging well. Moreover, each component includes further subparts. For example, active engagement with life comprises interpersonal relations and productive activity. Interpersonal relations encompass information exchange, transactions with others, emotional support, and direct assistance. Productive activity implies that an action should have social value, regardless if it is paid or not (e.g., taking care of grandchildren, taking care of ill relatives, volunteer actions, etc.).

Even if controversial, the model of successful aging still dominates social gerontology today (Liang and Luo 2012; Pruchno et al. 2010). Its popularity is linked to the fact that it was the first to raise awareness concerning the factors that may influence a good old age and promoted a new, positive approach to aging (Martinson and Berridge 2014). The *successful aging* concept effectively illustrates the aging paradox since it combines the concept of "aging," which had been previously associated with loss and dependency, with the notion of "success," signifying achievement, and productivity. Thus, the emergence of the successful aging model provided an opportunity to change negative views on aging that linked old age with illness and decline. By advocating the idea that aging can be associated with achievement, health, strength, and productivity, the successful aging model promoted positive views on aging. For example, images of active and happy old individuals are inspired by theories of successful aging. Such positive images of aging show what a desirable old age could be like and thus aim to make old age more attractive. However, the increase in positive depictions of older people in advertisements is also considered to be a consequence of the aging population becoming a more important segment on the market (Prieler et al. 2010, 2014, 2017).

Apart from the demographic changes that require a new way of thinking about aging and seniors, there are several other reasons why positive views on aging are relevant. In the following, I will discuss views on aging and their importance for health and well-being.

How Views on Aging Affect Our Lives

Views on aging represent beliefs about old age that people hold as individuals (views about one's own aging, also called self-perceptions of aging) or share as a social group (i.e., social cognitions of aging or stereotypes about aging). Positive views on aging comprise the hopes people form regarding their old age. They are defined as perceiving more gains than losses in connection to old age (Wurm et al. 2007, 2008). Thus, positive views on aging imply seeing potential for growth and development in old age and taking concrete actions to prevent physical, mental, and social decline. Considering aging to be positive infers that a person finds meaning in aging and perceives an ample time frame that motivates making plans and setting goals for old age (Wurm et al. 2010). By contrast, negative images of aging include perceiving a shorter subjective time frame, a strong focus on the present and undesirable prospects for old age such as increased illness and disability (Wurm et al. 2013).

Effects of Positive Views Versus Negative Views on Aging

Views on aging, when positive, can represent a resource for a happy old age. In contrast, negative views of aging may constitute an obstacle to good health in old age or a longer life in general. According to the existing scientific literature, there are three main pathways that explain how self-perceptions of aging may influence health: (1) physiological processes (Levy et al. 2000), (2) behavioral patterns (e.g., engagement in physical activity, Wurm et al. 2010), and (3) psychological mechanisms (e.g., perceived control, Levy et al. 2002).

Physiological or biological mechanisms are concerned with the effect of views on aging on the autonomic nervous system and its responses to environmental stress. For example, in an experiment where older adults were exposed to subliminal negative aging stereotypes, they experienced heightened cardiovascular response during stressful mathematical and verbal challenges (Levy et al. 2000). In contrast, participants who were exposed to positive aging stereotypes experienced reduced cardiovascular responses to stress (Levy et al. 2000). People are often exposed to stress in their lives and thus having negative views on aging constitute a vulnerability factor for repeatedly experiencing heightened cardiovascular responses to challenging situations. Such recurrent stress responses can lead to developing cardiovascular problems in old age (Levy 2009). Moreover, negative aging stereotypes were shown to inhibit recovery process after cardiovascular events (Levy et al. 2006).

Psychological mechanisms comprise expectations about aging that in time can become self-fulfilling prophecies (Levy 2009). Positive views on aging hold the potential to become self-fulfilling prophecies because they imply a higher control or self-efficacy beliefs (Levy et al. 2002). Evidence was found for a stereotype matching effect when using a priming experimental task (Levy and Leifheit-Limson 2009). This means that the positive or negative aging stereotypes was greatest when the content of the stereotypes (e.g., positive-cognitive such as "wisdom" or negative-cognitive like "senile"; positive-physical such as "lively" and negative-physical like "frail") corresponded to the outcome domains (i.e., either physical or psychological). Those study participants who were exposed to positive stereotypes outperformed those who were exposed to negative aging stereotypes in both cognitive and physical tasks (Levy and Leifheit-Limson 2009).

Behavioral pathways include engagement in either an unhealthier or healthier lifestyle. For example, performing little physical exercise is linked to negative beliefs regarding reduced capabilities when reaching old age. Negative age stereotypes imply that old age is associated with illness and thus physical workout becomes pointless (Levy and Myers 2004). Self-efficacy toward health behavior is also reduced by negative aging stereotypes (Levy et al. 2000). Sarkisian and colleagues (2002, 2005) found that people with negative views on aging were less likely to exercise and go to the doctor. On the contrary, having positive views on aging can give one confidence to exercise despite growing older and were found to be associated with engagement in health practices such as taking medication regularly (Levy and Myers 2004). Also, Wurm and colleagues (2010) showed that people with more positive views of aging engaged in more physical activity.

Negative images of aging have been shown to act as self-fulfilling prophecies concerning health in old age. Different researchers have found that people who hold negative expectations about old age tend to experience more aging-related health problems (Wurm et al. 2013; Levy et al. 2009). One possible explanation is that people who perceive a limited time frame will engage less in health behaviors (Wurm et al. 2010). Another explanation is that people who associate aging with illness consider it normal to experience health problems in old age and do little to prevent or treat such health issues. If people perceive aging as causing illness, then they believe that such illness are less preventable, curable, or controllable (Stewart et al. 2012; Sarkisian et al. 2001). Furthermore, Wurm and colleagues (2013) argue that as people grow older, negative perceptions of aging influence the use of selection, optimization, and compensation self-regulatory strategies (SOC, Baltes and Baltes 1990). Specifically, individuals who hold negative images of aging are less likely to use SOC strategies after experiencing a serious health event (Wurm et al. 2013). This may have a negative effect on their health (Stewart et al. 2012). One explanation is that negative images of aging imply that people attribute health decline to a stable, internal, global, uncontrollable cause, namely aging.

By contrast, there is a solid body of evidence that argues positive views on aging represent a valuable resource for health (Wolff et al. 2016; Wurm et al. 2007, 2010). Such studies show that self-perceptions of aging are associated with physical health (Wurm et al. 2007), with mortality (Kotter-Grühn et al. 2009; Levy and Myers 2004), physical functioning (Sargent-Cox et al. 2012), self-rated health, and life satisfaction (Wurm et al. 2008). However, some authors have asked themselves whether a good health determines more positive views on aging rather than views on aging influencing health. Findings from such studies suggest that self-perceptions of aging have a greater impact on health than the other way around (Sargent-Cox et al. 2012; Wurm et al. 2007) and establish their importance as an asset for positive aging.

Should We Think Differently About Aging? Critiques to Positive Aging Theory

Despite being one of the main mainstream models within social gerontology, successful aging theory has attracted a lot of critiques (Martinson and Berrige 2014; Liang and Luo 2012). In the next paragraphs, I will summarize some critical points and discuss how views on aging can change along with potential advances in developing the successful aging theory. Making this link is relevant since we can change our views on aging by modifying the theories that attempt to explain how we can age well.

Life Course Perspective and Process Models Versus Outcomes

According to the successful aging model, in order to age successfully, a person needs to meet a clear set of criteria. Nevertheless, studies pointed out that only a small percentage of individuals who are older than 65 meet the successful aging standards (McLaughlin et al. 2009; Hank 2011). Thus, successful aging criteria were criticized for being rigid, normative, and unrealistic (Martinson and Berridge 2014). Baltes and Carstensen (1996) argued that Rowe and Kahn's model disregarded the variety of possibilities for aging well and ignored the dynamic of aging. Furthermore, apart from setting unrealistic standards for successful aging and describing those who would qualify as aging successfully, the model failed to explain how a person may reach the desired old age. The question of what factors influence successful aging remained open for discussion.

Concerned with the narrowness of successful aging standards, some researchers proposed adding further criteria to the model such as physiological, social, and psychological dimensions (Young et al. 2009). A series of studies followed, aiming to find missing components of the successful aging model. Consequently, additional criteria were proposed including: spirituality (Crowther et al. 2002), marital status, and satisfaction (Ko et al. 2007), positive health (Kaplan et al. 2008), or a fusion with the generativity concept (Villar 2011). Ryff (1989) suggested that successful aging is linked to a series of psychological factors such as self-acceptance, meaning in life, environmental mastery, personal growth, autonomy, and positive social relations. However, despite the multitude of identified factors, the relations between factors and the mechanisms leading to a good old age remained unexplained.

As a response, the latest generation of resource models focuses on dynamic processes and considers positive aging as development across the entire life span (Stowe and Cooney 2014). The most empirically tested models of this category are the dual-processing model (Brandtstädter 1999), the Selection, Optimization and Compensation (SOC, Baltes and Baltes 1990) model, and the life span theory of control (Heckhausen and Schulz 1995).

The ***Selection, Optimization, and Compensation Model*** (Baltes and Baltes 1990) was developed as a process-oriented approach. It was formulated in an attempt to outline the processes involved in successful aging and to explain how people can reach a happy old age. Baltes and Baltes (1990) defined development in old age as maintaining a positive balance between gains and losses. In their view, people select from the life domains that are important to them and form goals (selection), optimize their resources and means of reaching the goals that are still important to them (optimization), and compensate for losses in the specific domains they have chosen (compensation) in order to ensure a lifelong successful development. The SOC processes occur either at a conscious or unconscious level and aim at maximizing gains and minimizing losses while striving to reach personal goals (Freund et al. 1999). Although SOC strategies are considered part of a universal mechanism, they still depend on the individual and social context. For example, goals vary from person to person and they also are culturally and historically different (Baltes and Carstensen 2003). Thus, the SOC model allows for individual, non-normative courses of development in old age.

The SOC was integrated with the action-theoretical framework, in the sense that the application of SOC strategies is useful for attaining goals in life (Freund and Baltes 2002, 1998). This assumes that people set goals all their lives and must select between the goals they set since resources for attaining these goals are limited. Elective selection represents the process of choosing between goals, organizing goals in a coherent hierarchy, and committing to the goals they chose. Optimization refers to the process of acquiring the needed skills for goal attainment. When resources deteriorate or are lost due to aging, compensation becomes significant for reaching one's goals (e.g., one can ask for help, one can practice learning new skills, etc.). In later life, losses begin to outweigh gains. As a result, individuals focus on their most important goals, reconstruct their goal hierarchy, disengage from original goals, and set new goals.

In response to the critique that the successful aging model focuses only on static outcomes rather than on aging processes, the lifetime approach to aging was proposed. Jopp and colleagues (2006, 2010) claim that successful development across the life span requires a fruitful interplay between resources, strategies, and beliefs. In agreement, Freund (2008) argues that successful aging happens over the entire life span. Thus, there is no particular point in life when one starts to age successfully or unsuccessfully (Freund and Riediger 2003). Freund and Riediger (2001) define resources as means to achieving ones' goals. However, resources are important all over the life span because they guarantee successful aging. Just having resources (e.g., psychological, financial, biological, etc.) is not enough for reaching ones' goals. A person needs to invest and manage these resources wisely to increase the resource-based gains and decrease potential losses (Baltes and Baltes 1990). Managing resources is relevant at any age but becomes increasingly relevant with aging. Thus, successful aging is conceptualized as representing a fruitful management of resources (Freund 2008). The self-management theory (Steverink et al. 1998) claims that age-associated changes in resources and goals need to be considered when studying positive aging. Substitution and compensation of relevant goals are presented as the key to successful adaptation across the life span.

As I have discussed in the previous section, views on aging represent a relevant resource for health (Wolff et al. 2016) and often influence health behavior (Wurm et al. 2013, 2010). The role of preventive behavior for successful aging has been acknowledged by several researchers (Wolff et al. 2016; Wurm et al. 2010, 2013). Preventive behavior includes taking action to ensure future health and helping others to develop social networks for old age. In this sense, preventive actions constitute a form of proactive coping as one builds resources in the present to deal with future obstacles.

Proactive coping was proposed as an addition to the SOC strategies for old age preparation (Ouwehand et al. 2007). The authors argue that SOC strategies are usually directed at how people react to losses, while proactive coping represents a strategy for preventing future losses. Proactive coping may help maintain resources needed for optimization and compensation strategies, while avoiding disengagement from relevant goals (Ouwehand et al. 2007). Individuals are encouraged to plan for old age, reflect on their goal hierarchy, optimize their skills, compensate when resources are scarce, engage in preventive health behaviors, and create social resources. For example, volunteering was shown to be good for ones' health since it provides meaning in life for older people (Konrath et al. 2012). Thus, from a SOC perspective, a positive image of old age is embodied by dynamic, engaged older individuals who actively manage their resources to age well. In line with this approach, the middle-aged person should already be involved in proactive coping and thus adopt health behavior and plan for old age. Those who fail to do so contribute to the development of a negative image of aging individuals who are not responsible enough to plan their old age in advance or who engage in risk behavior. This is connected to a social image of old age and the value attached to the elderly within a certain society. A negative image of aging implies that older people may become a burden for their families (if they have one) or to society in general in old age.

In relation to process theories, potential views on aging include individuals who cope with losses and stay positive despite confronting aging specific decline (e.g., deteriorating health, loss of physical attractivity, etc.). Nevertheless, it is difficult to develop views on aging based only on successful aging values. Successful aging criteria define certain standards (e.g., for biological, psychological or social functioning) or may motivate positive aging scenarios. Nevertheless, positive aging strategies might influence views on aging because they inspire people to picture themselves as happy older individuals or imagine themselves performing actions that can lead to a happy old age. Process-oriented views on aging would help people imagine what they can do to age successfully and picture themselves performing actions to reach a happy old age.

Success as a Controversial Value

The "hard hitting" critiques described by Martinson and Berridge (2014) condemned the core assumptions of the successful aging paradigm, namely the idea of "success" as an absolute standard for a good old age. Successful aging theory was criticized for using the notion of success as the main criteria to evaluate happiness in old age. From a critical gerontology perspective, "success" represents a neoliberal, Western value (Holstein and Minkler 2003) and its application leads to discrimination of those individuals who do not age according to standard. Critical gerontologists argued against the individualism promoted by the model, its neoliberal assumptions, its implicit ageism, its moral implications about the responsibility for aging well and potential for blaming those who do not manage to be healthy in old age. Furthermore, in line with this critique, the successful aging model was accused of promoting "agelessness," of encouraging an extension of youthful values and unrealistic young appearance in old age. By promoting a dualism between body and mind (i.e., not feeling as old as one is physically, not feeling one's chronological age), "successful aging" may lead to self-hatred and self-denial (Andrews 1999). Thus, it was argued that successful aging discourses represent a masked form of ageism. The discourse further discriminates by gender (Calasanti and Slevin 2001) and social class since not all individuals have equal resources for aging well. For example, the Dove campaign using older women as models was criticized for promoting the male gaze on the female body and promoting unrealistic standards of beauty (Craciun 2018). Overall, successful aging discourses may promote social and health inequalities in old age.

The "busy ethic" ethos that underlies successful aging was especially critiqued. According to critical gerontology, from a successful aging point of view, older people are only valued if they contribute to the economy, they either produce or are good consumers in their old age (Minkler and Holstein 2008). This has been interpreted as a marketing maneuver to promote an anti-aging industry (Moody 2005), described as "*the new sex for old*" (Katz and Marshall 2003). Successful aging becomes an individual choice, and the person is responsible for success or failure to age well. Views on aging that are inspired by success as an ultimate value imply that individuals are making independent choices, are being active (i.e., busy bodies) and contribute to society through productive activities such as work or volunteering.

Following the critique that success represents a Western value and may not be relevant in other cultures, researchers looked for alternative ideals for a happy old age. Nevertheless, few researchers attempted to formulate alternative frameworks. Liang and Luo (2012) argue for a shift in gerontological thought from successful to ***harmonious aging***. In line with the critical gerontology arguments, Liang and Luo (2012) criticize the successful aging model for supporting agelessness rather than positive aging, encouraging a "busy body" ethic, promoting capitalistic ideas of productive elderly and for being culturally blind. First, the authors argue that successful aging denies the changes inherent to aging and promote a cultural ideal of agelessness (Andrews 1999). This ideal can determine self-denial and self-hatred and a disharmony between body and soul. Second, the "busy ethic" and the importance of quantity of activities in old age overlooks the importance of activity quality in old age (i.e., is it important if people engage in many activities or they engage in less, but these are the activities that they cherish). Third, successful aging is considered to stand for capitalistic values, portraying older adults as consumers of lifestyles. Forth, the idea of success is considered to be cultural. Success portrays Western especially US values and may not be characteristic of other cultures. Liang and Luo (2012) propose the harmonious aging model as an alternative to successful aging. The idea of harmonious aging is inspired by Eastern philosophy. Harmony emphasizes the importance of maintaining a balance in all spheres of life, based on differences as opposed to uniformity. Body and mind are indispensable for health and act as a whole. Harmony also highlights human relations as well as a connection and interdependence between generations.

Views on aging are dependent on cultural contexts as well as on social contexts (Wolff et al. 2018; Fung 2013). Even though some characteristics may be universal, the way individuals picture old age may differ according to culture and social background. Existing cultural values are internalized by aging individuals in a process of cultural learning (Vygotsky 1934/1962) and become goals that guide their adult development. Thus, when individuals from different cultures pursue their aging goals, differences in socioemotional aging happen (Fung 2013). Images of aging are shared in specific contexts as cultural values and therefore people internalize views on aging from their own culture (e.g., showing respect for elderly, associating old age with wisdom, the duty of taking care of older relatives, etc.) as they age.

Views on aging inspired by this harmonious aging perspective may embody other meanings of success or other values that represent a positive old age. For example, instead of "busy bodies," old people may be represented as resting or relaxing or being engaged in contemplation such as in the Hindi culture (Ramamurti and Jamuna 2010). Different notions of beauty and happiness in old age may be considered. Thus, cultural differences can serve to inspire novel positive views on aging that do not necessarily imply being successful.

Missing Voices- Lay People Perspectives on Successful Aging

A further critique addressed to the successful aging model concerns the fact that it is ignoring subjective definitions of successful aging formulated by old adults themselves, namely the so-called missing voices (Martinson and Berrige 2014). Studies comparing theory driven definitions with lay definitions of aging found that usually the former do not cover the variability of options for positive aging that the latter comprise. Several components should be added to the successful model as for example spirituality (Lewis 2011), being open to change (Rossen et al. 2008), self-care, aging acceptance and financial well-being (Hilton et al. 2012), and emotional care received from family (Hsu 2007). Some researchers concluded that lay persons definitions of successful aging are relatively consistent over time and should be taken into account (Tate et al. 2013).

Lay perspectives on positive aging have been investigated less often than theory driven approaches (Jopp et al. 2014). On the one hand, lay perspectives are relevant for health policies and interventions since those policies and programs that correspond to lay people views will also get their involvement and commitment. On the other hand, lay perspectives on positive aging can further inform theory formation in the domain of positive aging.

In line with the "missing voices" critique, cultural relevance and variability constitute missing points in the successful aging paradigm. Authors have argued for taking subjective views from diverse cultural perspectives into consideration when studying positive aging. As discussed above concerning cultural views on aging, several authors have criticized the Western, white, and middle-class conceptualizations of successful aging by pointing out its limited cross-cultural relevance (Liang and Luo 2012; Iwamasa and Iwasaki 2011; Ng et al. 2011; Torres 2006, 2001). Consequently, a research field emerged concerning the examination of cultural differences in lay persons' views on successful aging (Fernández-Ballesteros et al. 2008, 2010; Hilton et al. 2012). Differences were usually found between Eastern and Western cultures. All in all, findings from such intercultural studies suggest that people perceive successful aging as a combination of basic resources such as health or finances, psychological factors such as cognitive and behavioral strategies (goals, activities, etc.) and beliefs (e.g., meaning in old age, positive attitudes). For example, Jopp and colleagues (2015) showed differences between US and Germany in what concerns views on aging. For Americans, self-responsibility in shaping aging is much more important than for Europeans. Also, they are more likely to experience old age as growth than Europeans do (Jopp et al. 2014; Westerhof et al. 2012).

How a certain society deals with its elderly or how older persons in a social context perceive their aging depends on the images of aging existing in that particular social or cultural context. Images of aging differ across time and space. However, in the age of globalization, social differences (e.g., how the social system is organized in a country) may be more important than cultural differences for shaping views on aging. Views on aging do not depend only on cultural contexts. Previous results have shown that people with high education tend to have more positive images of aging (Wurm and Huxhold 2012). However, in case of people who are living in precarious circumstances, characterized by risk and insecurity, even if they are highly educated, they may have more negative images of their old age because they feel vulnerable when facing age-related changes. Nevertheless, they may also hold certain positive views on aging, create novel views on aging, or manage to find a balance between negative and positive aspects of old age. Thus, research was needed to investigate this matter of views on aging in precarious individuals in more detail (Craciun and Flick 2014).

Images of aging depend on the social context also in the sense that distinctive professional categories may foster various images of aging (Berner et al. 2012). For instance, doctors who work directly with older individuals may hold different images of aging compared to IT programmers who have less contact with older people. There is a growing body of literature on images of aging in health professionals (Walter 2006; Samra et al. 2015; Craciun 2016; Leung et al. 2011) as well as how patients perceive the doctors' attitude toward them (MacRae 2016a, b). Nevertheless, views on aging may differ not only according to the profession (e.g., Health professional or IT specialist), but according to the nature of the job (i.e., secure job contract or precarious work). A social context characterized by precarity may inspire specific views on aging that might be either negative or positive and differ from the views on aging that emerge in a financially secure social context. Consequently, one needs to ask how would views on aging "look like" in a context of precarity and what roles would old people play in a precarious society? Following this line of argument, one of the main research questions of the present book was how does precarity shape views on aging? Potential answers to this question will be explored in the next chapters (see chapter six to ten). However, before discussing the findings of the research, in the next two chapters I will analyze the existing links between active aging policy and views on aging as well as between precarity as a social reality and views on aging.

References

Andrews, M. (1999). The seductiveness of agelessness. *Ageing & Society, 19*(3), 301–318. https://doi.org/10.1017/s0144686x99007369.

Atchley, R. (1999). *Continuity and adaptation in aging*. Baltimore, MD: The Johns Hopkins University Press.

Baltes, P. B., & Baltes, M. M. (Eds.). (1990). *Successful aging*. Cambridge University Press. https://doi.org/10.1017/cbo9780511665684.

Baltes, P. B., & Baltes, M. M. (1991). *Successful aging: Perspectives from the behavioral sciences*. Cambridge: Cambridge University Press.

Baltes, M. M., & Carstensen, L. L. (1996). The process of successful ageing. *Ageing and Society, 16*(4), 397–422. https://doi.org/10.1017/s0144686x00003603.

Baltes, M. M., & Carstensen, L. L. (2003). The process of successful aging: Selection, optimization, and compensation. *Understanding Human Development*, 81–104. https://doi.org/10.1007/978-1-4615-0357-6_5.

Bowling, A. (2005). What is successful aging and who should define it?. *British Medical Journal, 331*, 1548–1551.

Bowling, A. (2007). Aspirations for older age in the 21st century: What is successful aging? *International Journal of Aging and Human Develop-ment, 64*, 263–297.

Bowling, A., & Dieppe, P. (2005). What is successful ageing and who should define it? *British Medical Journal, 331*, 24–31.

Berner, F., Rossow, J., & Schwitzer, K.-P. (Eds.). (2012). Altersbilder in der Wirtschaft, im Gesundheitswesen und in der pflegerischen Versorgung [Views on aging in the industry, in health and care settings]. https://doi.org/10.1007/978-3-531-93287-3.

Brandtstädter, J. (1999). The self in action and development: Cultural, biosocial, and ontogenetic bases of intentional self-development. In J. Brandtstädter & R. M. Lerner (Eds.), *Action and*

self-development: Theory and research throughout the life span (pp. 37–65). Thousand Oaks, CA: Sage.
Calasanti, T. M., & Slevin, K. F. (2001). *Gender, social inequalities, and aging*. Walnut Creek, CA: Alta Mira Press.
Carstensen, L. L., Isaacowitz, D. M., & Charles, S. T. (1999). Taking time seriously: A theory of socioemotional selectivity. *American Psychologist, 54,* 165–181.
Crăciun, I. C. (2018). Aging in precarious times. *Alter Und Geschlecht*, 163–175. https://doi.org/10.1007/978-3-658-20872-1_11.
Craciun, C. (2016). "They Do Not Know How To Use Their Health Insurance Card": Implementation of active aging policy and aging perceptions in romanian general practitioners. *Ageing International, 41*(3), 254–264. https://doi.org/10.1007/s12126-015-9239-x.
Craciun, C., & Flick, U. (2014). "I will never be the granny with rosy cheeks": Perceptions of aging in precarious and financially secure middle-aged Germans. *Journal of Aging Studies, 29,* 78–87. https://doi.org/10.1016/j.jaging.2014.01.003.
Crowther, M., Parker, M., Achenbaum, W., Larimore, W., & Koenig, H. (2002). Rowe and Kahn's model of successful aging revisited: Positive spirituality—the forgotten factor. *The Gerontologist, 42,* 613–620.
Cumming, E., & Henry, W. E. (1961). *Growing old: The process of disengagement*. New York: Basic Books.
Dovenmuehle, R. H., Busse, E. W., & Newman, G. (1970). Physical problems of old age. In E. Palmore (Ed.), *Normal aging: Reports from the Duke longitudinal Study, 1959–1969* (pp. 29–50). Durham, NC: Duke University Press.
Fernández-Ballesteros, R., García, L. F., Abarca, D., Blanc, L., Efklides, A., Kornfeld, R., … Patricia, S. (2008). Lay concept of aging well: Cross-cultural comparisons. *Journal of the American Geriatrics Society, 56,* 950–952. https://doi.org/10.1111/j.1532-5415.2008.01654.x.
Fernández-Ballesteros, R., García, L. F., Abarca, D., Blanc, E., Efklides, A., Moraitou, D., … Patricia, S. (2010). The concept of ageing well in Latin American and European countries. *Ageing and Society, 30,* 41–56. https://doi.org/10.1017/s0144686x09008587.
Freund, A. M. (2008). Successful aging as management of resources: The role of selection, optimization, and compensation. *Research in Human Development, 5*(2), 94–106. https://doi.org/10.1080/15427600802034827.
Freund, A. M., & Baltes, P. B. (1998). Selection, optimization, and compensation as strategies of life management: Correlations with subjective indicators of successful aging. *Psychology and Aging, 13,* 531–543. https://doi.org/10.1037/0882-7974.13.4.531.
Freund, A. M., Li, K. Z. H., & Baltes, P. B. (1999). Successful development and aging: The role of selection, optimization, and compensation. In J. Brandtstädter & R. M. Lerner (Eds.), *Action and self-development: Theory and research through the life span* (pp. 401–434). Thousand Oaks: Sage Publications.
Freund, A. M., & Riediger, M. (2001). What I have and what I do: The role of resource loss and gain throughout life. *Applied Psychology: An International Review, 50,* 370–380.
Freund, A. M., & Baltes, P. B. (2002). Life-management strategies of selection, optimization and compensation: Measurement by self-report and construct validity. *Journal of Personality and Social Psychology, 82*(4), 642–662. https://doi.org/10.1037/0022-3514.82.4.642.
Freund, A. M., & Riediger, M. (2003). Successful aging. In R. M. Lerner, M. A. Easter, & J. Mistry (Eds.), *Handbook of psychology* (Vol. 6, pp. 601–628)., Developmental psychology Hoboken, NJ: Wiley.
Fung, H. H. (2013). Aging in Culture. *The Gerontologist, 53*(3), 369–377. https://doi.org/10.1093/geront/gnt024.
Hank, K. (2011). How "successful" do older Europeans age? Findings from SHARE. *The Journals of Gerontology. Series B, Psychological Sciences and Social Sciences, 66B,* 230–236. https://doi.org/10.1093/geronb/gbq089.
Havighurst, R. J. (1961). Successful aging. *The Gerontologist, 1,* 8–13. https://doi.org/10.1093/geront/1.1.8.

Havighurst, R. J. (1963). Successful aging. In R. H. Williams, C. Tibbitts, & W. Donahue (Eds.), *Processes of aging* (pp. 299–320). New York: Atherton Press.

Heckhausen, J., & Schulz, R. (1995). A life-span theory of control. *Psychological Review, 102,* 284–304.

Hill, R. D. (2005). *Positive aging: A guide for mental health care professionals and consumers.* New York: W.W. Norton.

Hill, R. D. (2011). A positive aging framework for guiding geropsychology interventions. *Behavior Therapy, 42*(1), 66–77. https://doi.org/10.1016/j.beth.2010.04.006.

Hilton, J. M., Gonzalez, C. A., Saleh, M., Maitoza, R., & Anngela- Cole, L. (2012). Perceptions of successful aging among older Latinos, in cross-cultural context. *Journal of Cross-Cultural Gerontology, 27,* 183–199. https://doi.org/10.1007/s10823-012-9171-4.

Holstein, M. B., & Minkler, M. (2003). Self, society, and the "New Gerontology". *The Gerontologist, 43*(6), 787–796.

Hsu, H. C. (2007). Exploring elderly people's perspectives on successful ageing in Taiwan. *Ageing & Society, 27,* 87–102. https://doi.org/10.1017/S0144686X06005137.

Iwamasa, G. Y., & Iwasaki, M. (2011). A new multidimensional model of successful aging: Perceptions of Japanese American older adults. *Journal of Cross-Cultural Gerontology, 26,* 261–278. https://doi.org/10.1007/s10823-011-9147-9.

Jopp, D., & Smith, J. (2006). Resources and life-management strategies as determinants of successful aging: On the protective effect of selection, optimization, and compensation. *Psychology and Aging, 21,* 253–265. https://doi.org/10.1037/0882-7974.21.2.253.

Jopp, D. S., & Schmitt, M. (2010). Dealing with negative life events: Differential effects of personal resources, coping strategies and control beliefs. *European Journal on Ageing, 7,* 167–180. https://doi.org/10.1007/s10433-010-0160-6.

Jopp, D. S., Wozniak, D., Damarin, A. K., De Feo, M., Jung, S., & Jeswani, S. (2014). How could lay perspectives on successful aging complement scientific theory? Findings from a U.S. and a german life-span sample. *The Gerontologist, 55*(1), 91–106. https://doi.org/10.1093/geront/gnu059.

Kaplan, M. S., Huguet, N., Orpana, H., Feeny, D., McFarland, B. H., & Ross, N. (2008). Prevalence and factors associated with thriving in older adulthood: A 10-year population based study. *The Journals of Gerontology. Series A, Biological Sciences and Medical Sciences, 63A,* 1097–1104. https://doi.org/10.1093/gerona/63.10.1097.

Katz, S., & Marshall, B. (2003). New sex for old: Lifestyle, consumerism, and the ethics of aging well. *Journal of Aging Studies, 17*(1), 3–16. https://doi.org/10.1016/s0890-4065(02)00086-5.

Ko, K. J., Berg, C. A., Butner, J., Uchino, B. N., & Smith, T. W. (2007). Profiles of successful aging in middle-aged and older adult married couples. *Psychology and Aging, 22,* 705–718. https://doi.org/10.1037/0882-7974.22.4.705.

Konrath, S., Fuhrel-Forbis, A., Lou, A., & Brown, S. (2012). Motives for volunteering are associated with mortality risk in older adults. *Health Psychology, 31*(1), 87–96. https://doi.org/10.1037/a0025226.

Kotter-Grühn, D., Kleinspehn-Ammerlahn, A., Gerstorf, D., & Smith, J. (2009). Self-perceptions of aging predict mortality and change with approaching death: 16-year longitudinal results from the Berlin Aging Study. *Psychology and Aging, 24,* 654–667. https://doi.org/10.1037/a0016510.

Kunzman, U., Little, T. D., & Smith, J. (2000). Is age-related stability of subjective well-being a paradox? Cross-sectional and longitudinal evidence from the Berlin Aging Study. *Psychology and Aging, 15,* 511–526.

Levy, B. R., Hausdorff, J. M., Hencke, R., & Wei, J. Y. (2000). Reducing cardiovascular stress with positive self-stereotypes of aging. *Journals of Gerontology. Series B, Psychological Sciences and Social Sciences, 55,* P205–P213.

Levy, B. R.; Slade, M. D., & Kasl, S. V. (2002): Longitudinal benefit of positive self-perceptions of aging on functional health. In: *Journal of Gerontology, 57*(5), 409–417.

Levy, B. R., & Myers, L. M. (2004). Preventive health behaviors influenced by self-perceptions of aging. *Preventive Medicine, 39*(3), 625–629. https://doi.org/10.1016/j.ypmed.2004.02.029.

Levy, B. R., Slade, M. D., May, J., & Caracciolo, E. A. (2006). Physical recovery after acute myocardial infarction: Positive age selfstereotypes as a resource. *International Journal of Aging and Human Development, 62,* 285–301.
Levy, B. (2009). Stereotype Embodiment. *Current Directions in Psychological Science, 18*(6), 332–336. https://doi.org/10.1111/j.1467-8721.2009.01662.x.
Levy, B. R., & Leifheit-Limson, E. (2009). The stereotype matching effect: Greater influence on functioning when age stereotypes correspond to outcomes. *Psychology and Aging, 24,* 230–233.
Levy, B. R., Zonderman, A., Slade, M. D., & Ferrucci, L. (2009). Negative age stereotypes held earlier in life predict cardiovascular events in later life. *Psychological Science, 20,* 296–298.
Leung, S., Lo, Giudice D., Schwarz, J., & Brand, C. (2011). Hospital doctors' attitudes towards older people. *Internal Medicine Journal, 41*(4), 308–314.
Lewis, J. P. (2011). Successful aging through the eyes of Alaska Native elders. What it means to be an elder in Bristol Bay, AK. *The Gerontologist, 51,* 540–549. https://doi.org/10.1093/geront/gnr006.
Liang, J., & Luo, B. (2012). Toward a discourse shift in social gerontology: From successful aging to harmonious aging. *Journal of Aging Studies, 26*(3), 327–334. https://doi.org/10.1016/j.jaging.2012.03.001.
Löckenhoff, C. E., & Carstensen, L. L. (2004). Socioemotional selectivity theory, aging, and health: The increasingly delicate balance between regulating emotions and making tough choices. *Journal of Personality, 72*(6), 1395–1424. https://doi.org/10.1111/j.1467-6494.2004.00301.x.
MacRae, H. (2016a). "My opinion is that doctors prefer younger people": Older women, physicians and ageism. *Ageing & Society, 38*(02), 240–266. https://doi.org/10.1017/s0144686x16001021.
MacRae, H. (2016b). "It's my body, my future": Older women's views of their interactions with physicians. *Journal of Women & Aging, 28*(3), 211–224. https://doi.org/10.1080/08952841.2014.950910.
Martinson, M., & Berridge, C. (2014). Successful aging and its discontents: A systematic review of the social gerontology literature. *The Gerontologist, 55*(1), 58–69. https://doi.org/10.1093/geront/gnu037.
McLaughlin, S. J., Connell, C. M., Heeringa, S. G., Li, L. W., & Roberts, J. S. (2009). Successful aging in the United States: Prevalence estimates from a national sample of older adults. *The Journals of Gerontology, Series B: Psychological Sciences and Social Sciences, 65B*(2), 216–226. https://doi.org/10.1093/geronb/gbp101.
Minkler, M., & Holstein, M. B. (2008). From civil rights to…civic engagement? Concerns of two older critical gerontologists about a "new social movement" and what it portends. *Journal of Aging Studies, 22*(2), 196–204.
Moody, H. R. (2005). From successful aging to conscious aging. In M. L. Wykle, P. J. Whitehouse, & D. L. Morris (Eds.), *Successful aging through the lifespan: Intergenerational issues in health* (pp. 55–68). New York, NY: Springer.
Ng, S., Cheung, C., Chong, A. L., Woo, J., Kwan, A. H., & Lai, S. (2011). Aging well socially through engagement with life: Adapting Rowe and Kahn's model of successful aging to Chinese cultural context. *International Journal of Aging and Human Development, 73,* 313–330. https://doi.org/10.2190/ag.73.4.c.
Ouwehand, C., de Ridder, D. T. D., & Bensing, J. M. (2007). A review of successful aging models: Proposing proactive coping as an important additional strategy. *Clinical Psychology Review, 27*(8), 873–884. https://doi.org/10.1016/j.cpr.2006.11.003.
Palmore, E. (1970). *Normal aging: Reports from the Duke longitudinal Study, 1959–1969* (pp. vii–ix). Durham, NC: Duke University Press.
Phillipson, C. (2013). *Ageing.* Cambridge: Polity Press. ISBN 13: 978 0 7456 3085 4.
Prieler, M., Kohlbacher, F., Hagiwara, S., & Arima, A. (2010). Silver advertising: Older people in Japanese TV Ads. *The Silver Market Phenomenon*, 239–247. https://doi.org/10.1007/978-3-642-14338-0_18.

Prieler, M., Kohlbacher, F., Hagiwara, S., & Arima, A. (2014). The representation of older people in television advertisements and social change: The case of Japan. *Ageing & Society, 35*(04), 865–887. https://doi.org/10.1017/s0144686x1400004x.
Prieler, M., Kohlbacher, F., Hagiwara, S., & Arima, A. (2017). Gender portrayals of older people in Japanese television advertisements. *Asian Women, 33*(1), 25. https://doi.org/10.14431/aw.2017.03.33.1.25.
Pruchno, R. A., Wilson-Genderson, M., Rose, M., & Cartwright, F. (2010). Successful aging: Early influences and contemporary characteristics. *The Gerontologist, 50,* 821–833. https://doi.org/10.1093/geront/gnq041.
Ramamurti, P. V., & Jamuna, D. (2010). Geropsychology in India. In G. Misra (Ed.), *Psychology in India: Clinical and health psychology* (Vol. 3, pp. 185–252). Noida, India: Dorling Kindersley.
Rossen, E. K., Knafi, K. A., & Flood, M. (2008). Older women's perceptions of successful aging. *Activities, Adaptation and Aging, 32,* 73–88. https://doi.org/10.1080/01924780802142644.
Rowe, J. W., & Kahn, R. L. (1987). Human aging: Usual and successful. *Science, 237,* 143–149.
Rowe, J. W., & Kahn, R. L. (1997). Successful aging. *The Gerontologist, 37,* 433–440.
Ryff, C. D. (1989). Beyond Ponce de Leon and life satisfaction: New directions in quest of successful ageing. *International Journal of Behavioural Development, 12,* 35–55.
Samra, R., Griffiths, A., Cox, T., Conroy, S., Gordon, A., & Gladman, J. R. F. (2015). Medical students' and doctors' attitudes towards older patients and their care in hospital settings: A conceptualisation. *Age & Aging, 44*(5), 776–783.
Sargent-Cox, K. A., Anstey, K. J., & Luszcz, M. A. (2012). The relationship between change in self-perceptions of aging and physical functioning in older adults. *Psychology and Aging, 27,* 750–760. https://doi.org/10.1037/a0027578.
Sarkisian, C. A., Hays, R. D., Berry, S. H., & Mangione, C. M. (2001). Expectations regarding aging among older adults and physicians who care for older adults. *Medical Care, 39*(9), 1025–1036. https://doi.org/10.1097/00005650-200109000-00012.
Sarkisian, C. A., Hays, R. D., & Mangione, C. M. (2002). Do older adults expect to age successfully? The association between expectations regarding aging and beliefs regarding healthcare seeking among older adults. *Journal of the American Geriatrics Society, 50,* 1837–1843. https://doi.org/10.1046/j.1532-5415.2002.50513.x.
Sarkisian, C. A., Prohaska, T. R., Wong, M. D., Hirsch, S., & Mangione, C. M. (2005). The relationship between expectations for aging and physical activity among older adults. *Journal of General Internal Medicine, 20,* 911–915. https://doi.org/10.1111/j.1525-1497.2005.0204.x.
Staudinger, U. M., Marsiske, M., & Baltes, P. B. (1993). Resilience and levels of reserve capacity in later adulthood: Perspectives from life-span theory. *Development and Psychopathology, 5*(04), 541. https://doi.org/10.1017/s0954579400006155.
Stewart, T. L., Chipperfield, J. G., Perry, R. P., & Weiner, B. (2012). Attributing illness to "old age": Consequences of a self-directed stereotype for health and mortality. *Psychology and Health, 27,* 881–897. https://doi.org/10.1080/08870446.2011.630735.
Steverink, N., Lindenberg, S., & Ormel, J. (1998). Towards understanding successful aging: Patterned change in resources and goals. *Ageing & Society, 18,* 441–467.
Stowe, J. D., & Cooney, T. M. (2014). Examining rowe and kahn's concept of successful aging: Importance of taking a life course perspective. *The Gerontologist, 55*(1), 43–50. https://doi.org/10.1093/geront/gnu055.
Tate, R. B., Swift, A. U., & Bayomi, D. J. (2013). Older men's lay definitions of successful aging over time: The Manitoba follow-up study. *International Journal of Aging and Human Development, 76,* 297–322. https://doi.org/10.2190/AG.76.4.b.
Torres, S. (2006). Different ways of understanding the construct of successful aging: Iranian immigrants speak about what aging well means to them. *Journal of Cross-Cultural Gerontology, 21,* 1–23. https://doi.org/10.1007/s10823-006-9017-z.
Torres, S. (2001). Understandings of successful ageing in the context of migration: The case of Iranian immigrants. *Ageing & Society, 21,* 333–355. https://doi.org/10.1017/S0144686X01008261.

Vaillant, G. E., & Mukamal, K. (2001). Successful aging. *American Journal of Psychiatry, 158,* 839–847. https://doi.org/10.1176/appi.ajp.158.6.839.

Villar, F. (2011). Successful ageing and development: The contribution of generativity in older age. *Ageing and Society, 32*(7), 1087–1105. https://doi.org/10.1017/s0144686x11000973.

Vygotsky, L. S. (1934/1962). *Thought and language*. Cambridge, MA: MIT Press.

Walter, U., Flick U., Neuber A., Fischer C., & Schwarz F.W. (2006). *Alt und gesund? Altersbilder und Präventionskonzepte in der ärztlichen und pflegerischen Praxis [Old and healthy? Views on aging and prevention from the health and care practice]*. Springer.

Westerhof, G. J., Whitbourne, S. K., & Freeman, G. P. (2012). The aging self in a cultural context: The relation of conceptions of aging to identity processes and self-esteem in the United States and the Netherlands. *The Journals of Gerontology, Series B: Psychological Sciences and Social Sciences, 67,* 52–60. https://doi.org/10.1093/geronb/gbr075.

Westerhof, G. J., Miche, M., Brothers, A. F., Barrett, A. E., Diehl, M., Montepare, J. M., ... Wurm, S. (2014). The influence of subjective aging on health and longevity: A meta-analysis of longitudinal data. *Psychology and Aging*, 29(4), 793–802. https://doi.org/10.1037/a0038016.

Wolff, J. K., Warner, L. M., Ziegelmann, J. P., Wurm, S., & Kliegel, M. (2016). Translating good intentions into physical activity: Older adults with low prospective memory ability profit from planning. *Journal of Behavioral Medicine, 39*(3), 472–482. https://doi.org/10.1007/s10865-015-9707-5.

Wolff, J. K., Wiest, M., Nowossadeck, S., Beyer, A., & Wurm, S. (2018). Regional impact of demographic change on changes in individual views on aging. *Innovation in Aging, 1*(suppl_1), 1335. https://doi.org/10.1093/geroni/igx004.4895.

Wurm, S., Tesch-Römer, C., & Tomasik, M. J. (2007). Longitudinal findings on aging-related cognitions, control beliefs, and health in later life. *The Journals of Gerontology Series B: Psychological Sciences and Social Sciences, 62*(3), P156–P164.

Wurm, S., Tomasik, M. J., & Tesch-Römer, C. (2008). Serious health events and their impact on changes in subjective health and life satisfaction: The role of age and a positive view on ageing. *European Journal of Ageing, 5,* 117–127. https://doi.org/10.1007/s10433-008-0077-5.

Wurm, S., Tomasik, M. J., & Tesch-Römer, C. (2010). On the importance of a positive view on ageing for physical exercise among middle-aged and older adults: Cross-sectional and longitudinal findings. *Psychology and Health, 25*(1), 25–42. https://doi.org/10.1080/08870440802311314.

Wurm, S., & Huxhold, O. (2012). Sozialer Wandel und individuelle Entwicklung von Altersbildern [Social change of individual views on aging]. In F. Berner et al. (Eds.) *Individuelle Und Kulturelle Altersbilder [Individual and cultural views on aging]*, 27–69. https://doi.org/10.1007/978-3-531-93286-6_2.

Wurm, S., Warner, L. M., Ziegelmann, J. P., Wolff, J. K., & Schüz, B. (2013). How do negative self-perceptions of aging become a self-fulfilling prophecy? *Psychology and Aging, 28*(4), 1088–1097. https://doi.org/10.1037/a0032845.

Young, Y., Frick, K. D., & Phelan, E. A. (2009). Can successful aging and chronic illness coexist in the same individual? A multidimensional concept of successful aging. *Journal of the American Medical Directors Association, 10,* 87–92. https://doi.org/10.1016/j.jamda.2008.11.003.

Chapter 3
The Politics of Positive Aging

Introduction

Aging happens in a social, cultural, and political context. Even if individuals can decide to a certain extent how they want to live their lives in old age, the decisions they make are shaped by policies (e.g., retirement, health, housing, etc.). Positive aging theories have influenced not only research but also policies concerning old age and aging individuals. In the following, I will discuss the policy of active aging and how it makes use of views on aging to promote health and well-being for the elderly population. In the present chapter, I analyze three case studies where changing views on aging is part of the positive aging policy agenda. The first refers to the activity of the World Health Organization (WHO) that seeks to promote health and well-being in old age by changing views on aging and lifestyles at individual level. The second case study refers to the European Union (EU) that aims to change views on aging to induce structural transformations that would make positive aging possible and sustainable. The third case study focuses on the German active aging policy that encourages the adoption of more positive views on aging to build a more age-friendly society and achieve healthier lifestyles. Germany was chosen because it represents an aging society with strong active aging policies. Moreover, Germany also makes the case study for the second section of the present book. Finally, some critiques of the active aging policy are summarized. The point is made that views on aging depend on the social reality where they are created or implemented. Thus, one needs to understand both the policy context and the social reality in order to make sense of aging experiences at individual level. The link is made to the next chapter where the social reality of aging and its potential discrepancies with the advertised views on aging are examined to a greater extent.

In the following, I will explore how aging policies evolved in parallel with variations in views on aging. The social context is known to shape views on aging (Levy 2009; Fernandez-Ballesteros et al. 2017). Views on aging in turn, may guide policy formation. Therefore, I will discuss how policy makes use of views on aging to change behavior at individual level as well as social structures. The starting point

I. C. Crăciun, *Positive Aging and Precarity*, International Perspectives on Aging 21, https://doi.org/10.1007/978-3-030-14255-1_3

for this second argument is the point made by Lassen and Moreira (2014) that active aging policy is "unmaking old age." In this sense, I argue that active aging policy uses views on aging as a tool to reform the way people think about old age and how they live their late life. I will further discuss the German active aging policy as a case study where the aim to change views on aging is formulated explicitly. In the concluding part of the chapter, I focus on critiques addressed to the active aging policy and the role played by views on aging in formulating these criticisms.

Defining Active Aging Policy: Do Views on Aging Play a Role?

Successful and positive aging ideas (see Chap. 2) inspired different institutions such as the World Health Organization (WHO) and the European Union (EU) to develop an **active aging policy** (Walker 2009; Foster and Walker 2014; Lassen and Moreira 2014). Within the active aging definition, the "aging" refers to the whole life course, while the "active" denotes ongoing development on several levels: social, economic, cultural, as well as daily life (Walker 2006). In contrast to the successful aging paradigm, active aging is said to be more holistic since it embraces a life-course-oriented approach (Foster and Walker 2014). Furthermore, it is considered it places at least in theory an emphasis on an individual as well as social responsibility for aging well (Foster and Walker 2014). The well-being of older people depends on the chances they are offered within a given social context (Villar 2012) such as continuous learning opportunities (Davey 2002), increasing employment prospects and reducing ageism at work (Corsi and Samek 2010; Walker and Maltby 2012), voluntary work possibilities (Deeming 2009).

According to WHO (2002), the **Active Aging Policy Framework** defines active aging as "*a process of maximizing opportunities for health, participation and security in order to enhance quality of life as people age*" (WHO 2002, pp. 12). The active aging policy framework is founded on three separate pillars. First, *the health pillar* includes policies that encourage independence and autonomy in old age and aim to reduce health care system costs. Second, *the participation pillar* targets the social sphere and seeks to improve access to social activities such as employment and education for older adults. The third pillar concerns the *security* of older individuals and aims to design actions that ensure protection, dignity, and care of the elderly.

Views on aging play an important part in the way active aging policy developed over time. On the one hand, the social context and existing aging stereotypes influence policy development. On the other hand, active aging policy attempts to change the social environment and peoples' behavior by changing views on aging. Thus, views on aging represent both an inspiration and a tool for aging policy.

There are close links between policy and views on aging. On the one hand, policy is influenced by existing aging stereotypes. On the other hand, policy creates innovative views on aging to inspire individuals concerning how to organize their lives and their old age. Taking these viewpoints into consideration, in the following I will first explore how social changes inspire policy development through the promotion of views on aging at social level. Second, I will focus on active aging policy and summarize results from studies that looked at how active policy ideals are reflected by lay peoples' views on aging.

Policy Development and Views on Aging

Discourses on aging and old people should be regarded in light of the social and historical contexts where they evolved (Dillaway and Byrnes 2009; Wolff et al. 2018; Fung 2013). Demographic changes marked by an increase in the aging population gave rise to public concern (Barusch 2013; Van den Heuvel 2012; Walker 2009). Even if at individual level, a longer life may represent a reason for joy, at policy level, a growing aging population would inspire rather negative views concerning older people. From a public policy point of view, not aging itself but the increasing number of older individuals and the proportion of elderly within the population were regarded as problematic (Walker 2009; Dillaway and Byrnes 2009). These economic and social threats caused by the growing population of older individuals were accompanied by certain negative views on aging and old people. For example, public discourse frequently described the rising number of older individuals as a "*gray tsunami*" (Barusch 2013). The association with a natural catastrophe implies that old people embody a destabilizing force that affects the health and pension systems. Similar, mass media images often depicted older people as "greedy geezers" with the implication that they take relevant resources away from the young (Phillipson 2013) or as passive recipients of pensions (Walker 2009). Such negative views concerning senior citizens could only reinforce existing negative aging stereotypes where old people were regarded as ill and dependent or as representing a burden to society in general.

The link between aging and policy was created in the "golden age" of the welfare state after the II World War (Walker 2009). Although the association with the welfare state helped raise the quality of life of senior citizens, it also contributed to portraying older adults as "dependent" recipients of support. Negative aging stereotypes were emphasized, associating old age with dependence and frailty. Such negative images of aging lingered even after the poverty level of seniors in Western European countries had decreased (Naegele and Walker 2009).

During the 1970s and 1980s, the financial consequences of population aging first came into focus as a matter of public concern. Subsequently, older persons were being described as the "*selfish welfare generation*" (Phillipson 2013), while aging came to be defined in terms of a "financial burden" for society. Ageist attitudes limited the social opportunities for older individuals, causing them to experience negative emotions and financial dependency. As a consequence, in 1968 Butler proposed the term "ageism" to describe discriminatory actions toward older people. Being a senior rights advocate, Butler (1975) stated that ageism should be treated as a social disease. Among other solutions "for curing this disease," he suggested fighting negative aging stereotypes and promoting productive aging by offering paid work and volunteer opportunities to older adults. In this context, the notion of "productive aging" developed. Productive aging meant that older adults should become agents of their own lives. Activity choices could be regulated by productive aging policy. At the same time, more opportunities to be active could influence public perception of older individuals (Bass and Caro 2001). Civic engagement involves individual action for the common good (Adler and Goggin 2005). In what active aging is concerned, civic engagement is often used to describe volunteer activities with benefits for both older people and communities. On the one hand, older people are encouraged to volunteer for the community to find meaning in old age and secure their social support networks. On the other hand, communities must gain from the contribution of older citizens (Morrow-Howell 2011). What productive aging, civic engagement, and active aging policy share is a positive view of aging since they all imply that old age can be a time of well-being, engagement, and contribution to society.

During the 1990s, anxieties concerning older individuals changed. Aging started to be depicted as just another crisis that characterizes the emerging "risk society" (Beck 1992). Older individuals were associated with other casualties such as climate change or terrorism and were thought to destabilize social and political institutions (Mullan 2004). The 2000s marked the merge of the "economic burden" and "cultural crisis" discourses and created a moral panic associated with the baby boomer generation reaching retirement age (Phillipson 2013). Such negative discourses were linked to predictions about the high numbers of people entering retirement far outnumbering the young employees that had to support them. The "crisis scenario" concerned the effect of growing aging populations on the pension system as well as the health system (Van den Heuvel 2012; Estes 2001). Older individuals were seen as happily enjoying life, but at the "expense" of younger generations. The representation of the elderly was modified from the "*deserving poor*," who are not able to work as they used to in their younger years and need protection from the welfare state to "*greedy geezers*" who do not want to work and therefore drain the State resources (Schultz 2001).

As a reaction to the "aging as crisis" discourse, solutions such as increasing the retirement age and encouraging older people to be productive in their late life (Dillaway and Byrnes 2009) were proposed. However, such solutions needed to be supported by more positive views on aging that would encourage older people to take charge of their lives and stay independent in old age. In this context, successful aging theories (Rowe and Kahn 1997) could provide novel positive views on aging since they promoted desirable aging ideals that could inspire aging persons to change their attitudes and behaviors. Furthermore, since the baby boomers' generation possessed buying power, the market needed to adapt to the needs of the "elderly consumer." Also, by setting success as a standard for a fulfilled old age, novel needs were generated for the aging individual. As the silver economy flourished, it promoted a variety of positive views on aging from the idea of "fighting old age" to "silver travelers" (Potratz et al. 2009). Positive aging views helped to encourage people to buy anti-aging products (e.g., cosmetics, medicine) or engage in age-combating behavior (Smirnova 2012; Andrews 1999). Thus, the creation of successful aging discourses paralleled policy efforts of reconstructing the image of the older adult (Estes 2001; Martinson and Minkler 2006). The old individual was now to be depicted as active and useful for society. The successful aging notion that initially developed as a critique to the welfare state was currently supported by neoliberal ideas of productivity and individual responsibility for risk (e.g., health risk or work risk). Therefore, positive views of "planning a successful old age," "investing old age with meaning," as well as portrayal of happy and active old people emerged. Initiators of successful aging models may not have foreseen that positive aging ideas would lead to policies that will eventually reduce the power of the welfare state and strengthen the negative stereotypes concerning old people.

Changes in the historical and social context further altered both views on aging and active aging policy. Globalization and the spread of neoliberal ideology (Estes and Phillipson 2002; Walker 2009; Phillipson 2013) and the birth of international organizations in the 1880s impacted aging policy and their correspondent representations of aging. Views on aging ceased to be just country specific and became more global since international organizations turned into agents that promoted them all over the world. Furthermore, the technology development helped to spread images of aging beyond country borders as well as made active aging policy known across countries.

In this context, Kohli (1985, 2007) introduced a novel interpretation of the life course as a social institution. In his view, the life course could not be regarded just as an individual phenomenon but constituted a feature of social structures. Following this line of argument, age became a category to be integrated into social institutions and social structures through policy (Dannefer 2012). The concept of *chronologization* was proposed (Kohli 2007), referring to the temporal aspects of life standardization that are integral to the institutionalized life course. Kohli (2007) argued that the institutionalized life course helps persons invest their lives with meaning and foster their mental health. This institutionalized structure provides individuals with a set of biographical orientations to organize their experience and plans and provides rules to help people organize, conduct, and give meaning to their lives (Kohli 2007).

Despite their popularity, successful aging ideas were often criticized (see Chap. 2). While the successful aging model was being challenged, policy adjustments were needed to fit the latest ideas concerning positive aging and adapt to the evolving social contexts. For example, one domain where demographic changes affected policy is that of work and more specifically retirement from work. Flexible work forms became more common and linear transitions from work to retirement became less frequent (Phillipson 2013; Phillipson et al. 2018; Loretto and Vickerstaff 2015). What had once been described as the three boxes of education, work, and retirement (Riley and Riley 1994) with age-graded role structures was confronted with the demands of a growing aging population. Consequently, novel work and retirement policies needed to be devised. Such new ways of thinking about retirement required innovative views of active older workers or active retirees who are independent and resourceful. Images of retired old persons tending to their gardens needed to be replaced with images of working older adults.

Taking all the above into consideration, positive alternatives inspired by successful aging theories influenced policy making (Schultz 2001; Dillaway and Byrnes 2009) and views on aging played a role in this process.

How Policy Ideas Are Reflected by Lay Peoples' Views on Aging

Since policy attempts to change how people think about aging and old age, a relevant question is whether they succeed? Are views on aging influenced by the dominant social discourses? Inspired by such questions, several studies explored the views of lay people on active aging or successful aging (Craciun 2011; Craciun and Flick 2014; Clarke and Warren 2007; Bowling 2006; Bowling and Gabriel 2004; Stenner et al. 2011; Swift and Tate 2013; Nosraty et al. 2015). In general, results from such studies showed that peoples' perceptions of active aging were similar to policy prescriptions. For example, findings pointed out that successful aging was associated with health, physical functioning, social and leisure activity, showing that individuals have internalized active aging ideas (Clarke and Warren 2007; Davey 2002; Hui-Chuan 2007; Stenner et al. 2011). Nevertheless, there were also contradictions between policy demands and people's experiences and expectations of aging.

Bowling (2008) found that active aging was associated with physical health and functioning, with leisure and social activities as well as adequacy of financial circumstances. Maintaining good health and functioning by exercising the body and mind were among the most frequent active aging goals named by participants (Bowling 2008). These are in accordance with the WHO goals of active aging (WHO 2002). Hilton and colleagues (2012) discovered that aging Latinos focused on maintaining a positive outlook, living in the present, enjoying a sense of community, deriving a sense of meaning from spirituality and family but also identified several financial worries related to old age. In a similar way, Clarke and Warren (2007) revealed the importance of the subjective component of active aging for lay participants. Their findings pointed out the relevance of the subjective temporal horizon, expressed by "living one day at a time" or "living in the moment" (Clarke and Warren 2007).

Results from another study (Stenner et al. 2011) point out that the subjective meaning of an activity is relevant to aging adults. Their outcomes showed that people associated old age with mental, physical, and social activity and highlighted the importance of autonomy in old age (Stenner et al. 2011). The quantity of activities is not as relevant as the quality people attach to it. Thus, Stenner and colleagues (2011) highlighted the relevance of the meaning of activity in the lives of older adults. In their opinion, views on aging may embody the multitude of activities that are relevant to older people (e.g., gardening, gossiping, reading the newspaper, cooking), even if these are not directly targeted by policy (e.g., sports, education, volunteer work). In line with the autonomy principle promoted by active aging policy, individuals should choose themselves what they understand as pleasant activity in late life and integrate these actions into their positive views on aging (Stenner et al. 2011).

The agentic capacity expressed by living by ones' own norms appeared to be very important. According to their results (Stenner et al. 2011), the agentic capacity is what makes an activity meaningful and relevant for an individual. Participants named the most common obstacles they faced when trying to be active in their old age and how they coped with these difficulties. Resembling the selection, optimization, and compensation principles (Baltes and Baltes 1990) or to proactive coping (Ouwehand et al. 2007) these findings pointed out that older individuals were not passive when facing obstacles but actively thought out solutions to their problems. Being old was associated with passivity, loss of interest and autonomy and not with chronological age itself (Stenner et al. 2011). Participation and activity as understood by lay people does not necessarily fit with the ideas promoted by the active aging policy. For example, people described ordinary activities such as gardening and gossiping as being important to them and did not necessarily consider work or voluntary activities as crucial for "being active." In a similar way, results from Bowling's study (2008) showed that lay perceptions of active aging did not include aspects such as productivity or empowerment (Bowling 2008).

How Policy Is Changing Views on Aging

The active aging policy aims to achieve its goals by challenging negative stereotypes about aging and changing perceptions of old age and the elderly. By changing views on aging, one can construct a novel social reality of aging to replace the one characterized by illness, disability, and dependency. Such aims can be achieved in different ways contingent on the institutions that propose or implement the policy. In a thought-provoking article, Lassen and Moreira (2014) described the pathways by which WHO and EU attempt to unmake old age. Namely the authors argued that both WHO and EU "unmake" old age through their policies even if their focus and instruments for achieving their aims differ (Lassen and Moreira 2014). In the following, I will discuss how the active aging policy is considered to have "reinvented old age" in order to reach its goals.

WHO and Changing Views on Aging at Individual Level

In an effort to promote healthy aging, the World Health Organization developed a policy framework focused on reducing the disease burden and increasing health-enhancing factors among the elderly (WHO 2002). It aimed to make sure that older people receive the necessary care and that they can maintain their health that guarantees their social functioning and social integration. "Activity" is regarded not just as being gainfully employed in old age, but by being engaged in leisure activities, being healthy and able to function and take part in social activities. Autonomy is an important concept in the WHO active aging policy an refers to the ability to control, cope, and make decisions about how to live ones' life according to ones' own rules and wishes (WHO 2002).

One important strength of the active aging policy designed by WHO (2002) is that it shifts the focus from employment and retirement policy to health and well-being. Furthermore, it stresses the relevance of the life course perspective and includes elements such as social participation and life satisfaction in old age. The **WHO active aging policies** place their emphasis on individual lifestyle change, namely on making healthy choices to stay fit in advanced age. Individuals are encouraged to engage in health behavior (e.g. practice sports, eat healthy, decrease alcohol consumption, etc.) to reduce chances of illness and disability in old age. Since WHO adopts a life course approach to healthy aging, illness prevention is supposed to start early in life. Also, because "old age" was replaced by a more appropriate terminology of "aging," preventive efforts should be seen as a dynamic task for the whole lifespan. Activity is regarded as healthy, while passivity is considered unhealthy (Lassen and Moreira 2014, p. 39). As a consequence, positive representations of aging portray a healthy and active older individual, engaged in sports or social interactions. On the contrary, passive behavior like napping or watching TV become part of an outdated or even negative image of old age. Views on aging are used as a means to achieve behavior change at individual level. Aging adults may come to judge what constitutes appropriate "active behavior" in old age based on policy defined standards.

EU and Changing Views on Aging at Community Level

At a European level, the growing aging population and estimated pension costs raised concern and justified the introduction of new aging policies. Several challenges were identified such as the decline in the working-age population, an increased need for care, the rise in pension expenditures, and the diversity among older people. This led to the Commission proposing guidelines for policies such as improving social protection, reversing trends of early retirement, improve health policies, taking a stand against age discrimination at work (Foster and Walker 2014). Despite the larger extent of proposed measures, those directed at the increase in employment among older individuals were highlighted (Walker 2009; Hamblin 2010). This focus was expressed in the aim to increase employment among those aged 55–64 to 50% by 2010 (Foster and Walker 2014). None of the EU countries managed to reach this goal (Zaidi and Zolyomi 2011).

European Union (EU) active aging policies seek to institutionalize late life through changes in structures and expectations regarding old age (Lassen and Moreira 2014, p. 40). The rising aging population of EU states constitutes a challenge for welfare states with their pension and health care systems. Consequently, EU attempts to implement population changes and reforms to redefine old age. In order to achieve this, the term "old age" was replaced with "aging." The latter expresses a dynamic life stage instead of simply referring to a static age group. Since the active aging framework adopts a life course perspective to aging, EU policies target not just the elderly, but address all age groups. The EU active aging policy became a coordination of pension reforms, age integration, and management, changes in attitudes toward early withdrawal from work and old age in general. The EU policy focuses on productive aging, where inactivity is considered unhealthy as well as unjust and asocial (Lassen and Moreira 2014). Therefore, pension reforms aim to expand working lives and discourage early retirement and create flexible retirement schemes (Hamblin 2010; Foster and Walker 2014). Such pension reforms help create positive expectations of an active old age. However, the prolongation of the working life can also raise anxieties concerning the ability to work and finding jobs or the stress of competing on the job market in an advanced age. Thus, changes in policy could influence views on aging by creating positive or negative expectations regarding the future.

To reach a good old age, individuals need to engage in a series of preparatory actions. While according to WHO, preparing for old age involves several healthy lifestyle choices, according to EU, it implies changing expectations about aging, working after retirement, or volunteering for social causes. Failure to take preparatory actions to achieve a healthy old age was attributed to a lack of self-regulatory process (e.g., lack of planning skills) or how people choose some goals over others as described by the Selection Optimization Compensation model (SOC, Baltes and Baltes 1990). Socioemotional selectivity theory (Carstensen 1992; Carstensen, Fung and Charles 2003) postulates that when people are confronted with the finitude of their lives, their attention may shift from future-oriented goals to emotionally relevant ones. Fung and Carstensen (2006) showed that when sociocultural events prime the fragility of life (i.e., September 11 or the annexation of Hong Kong to China), both young and old people are more motivated to derive meaning out of life and value emotionally meaningful relations over social contacts that might be useful in the future. The importance that people attribute to their goals changes as a function of perceived time (Carstensen et al. 2003). When the future is perceived as open-ended, people tend to focus on information-seeking goals, while when limited time perspectives are perceived (i.e., When being diagnosed with a terminal illness, when reaching old age, etc.), people tend to value emotionally meaningful goals (balancing emotional states or feeling needed by others). The perceived time is also a component of views on aging. Positive views on aging are associated with planning and longer time frames, while negative views on aging are linked to perceiving a short-time frame and not planning. Thus, views on aging become a policy tool for unmaking old age. Sometimes, policy directly targets changing views on aging to promote active aging. Such is the case of Germany that I will discuss in the following.

Active Aging Policy Targeting Changes in Views on Aging: The Case of Germany

Germany is one of the European countries with the highest percentage of people aged over 65. German women can expect to live until the age of 83, while German men until 78 years old (Mahne et al. 2017). As a response to such demographic developments and the influence of successful aging ideas, active aging policies were elaborated (Tesch-Römer and Wurm 2012). These policies take the viewpoint that a longer life implies growth and development in old age, while a growing aging population represents a resource rather than a burden for society. Both research and policy in Germany embraced the WHO (2002) definition of active aging as well as successful aging principles (Rowe and Kahn 1998). The SOC (Baltes and Baltes 1990) inspires interventions and policies for active aging because the selection of goals, optimization of opportunities for reaching goals and compensation as substitution of lost resources are regarded as central mechanisms for active aging. One further assumption of the active aging policy concerns the variability of aging experiences. Another central concept of active aging is that of *plasticity*. This implies that people can still change in their old age and training or interventions can stimulate development in late life (Tesch-Römer 2012; Singer et al. 2003). Therefore, healthy aging may be fostered through psychological interventions (Wolff et al. 2014).

Views on aging constitute one of the central targets of the active aging policy in Germany. Studies have shown that although images of aging are changing for the better, nevertheless negative images of aging prevail (Wurm and Huxhold 2012). The potential of views on aging to increase chances for good health and longevity has been proven in a series of studies (Wurm et al. 2007, 2010). Several interventions were developed to attempt changing images of aging and as a result foster health behavior in old age (Wolff et al. 2014; Kotter-Grühn 2015). Nevertheless, views on aging should not be overtly positive since by excluding the experience of loss in old age, they would prove to be unrealistic and backfire. Therefore, images of aging should include both potentials and risks that are associated with old age.

The German active aging policy implies that how people cope with the challenges of old age depends on what views on aging they hold. Such assumptions are based on research results showing the importance of views on aging on health (Wurm et al. 2007, 2010) and longevity (Kotter-Grühn 2015). In the context of these studies, views on aging were defined as individual or societal representations of old age, of aging as a process and of older individuals as a social group. These views on aging are social constructions because they emerge in a certain cultural context or society. Aging represents a meaning making process and culture provides the values that help make sense of the old age experience and set goals that guide personal development (Fung 2013). Studies point out differences in images of aging between cultures (Hilton et al. 2012; Löckenhoff et al. 2009) differences between images of aging of immigrants compared to native populations (Weintritt 2012; Bowling 2009) and transcultural differences. The latter refer to changes that happen when people grow old in cultures that are different from their own (Torres 2006, 2014). Furthermore, images of aging are also characteristic to certain life domains. For example, there are certain images of aging associated with the work world and others that emerge in hospitals or care institutions for older individuals.

Active aging policy in Germany aims to take into account the plurality of existing views on aging from the German society. Recommendations from the German active aging policy comprise securing an environment where older people can maintain their autonomy, respecting the heterogeneity of aging experiences, promoting positive images of aging in work contexts and thereby reduce ageist practices. Furthermore, views on aging are targeted in relation to education since studies have shown people with better education also have positive views on aging (Wurm and Huxhold 2012). Another intervention aim concerns images of aging promoted in the mass media (e.g. TV, social media, press) because these play a significant role in the development of social representations of old age and older individuals (Vickers 2007; Rozanova 2010; Horton et al. 2007; Prieler et al. 2014).

Views on aging are pursued also in the health context. At individual level, views on aging are of key importance for adopting a healthy lifestyle (Levy 2003). Moreover, the health personnel's' images of aging affect health communication and practice (Eliassen 2015). Previous studies have shown that doctors and carers display rather negative images of aging (Walter et al. 2006; Craciun and Flick 2015). Thus, changing images of aging in the medical and care contexts represent targets of policy intervention.

All in all, the active aging policy attempts to develop a new culture of aging in Germany, based on seeing old people as a resource for society. Active aging policy aims to make older people active members of society in order to help maintain their autonomy and feel useful in old age. Volunteering in old age has been shown to be beneficial for health and well-being (Konrath et al. 2012). As a consequence, volunteer activities are particularly encouraged by policy (Naegele 2016). As summarized above, policy recommendations in Germany attempt to combat ageism in several domains from work to health and rights of older individuals. In order to achieve its goals, the active aging policy needs to construct old age as a dynamic process and applies a life course perspective in what concerns policy recommendations and implementation. This begs the question: did these active aging policies succeed in changing people's views on aging? Are older people in Germany becoming more active and are they happier?

Inspired by the concern whether active aging policy discourse on productive aging had a negative or positive impact on aging individuals, one study set out to investigate the link between activation demands (expectations to stay mentally and physically fit) and involvement in former productive roles (Pavlova and Silbereisen 2011). Results showed that perceiving activation demands was associated with better adjustment among young–old Germans (aged 56–75) who were not working or volunteering (Pavlova and Silbereisen 2011). Nevertheless, expectations for active aging seemed to decrease with older age. The authors suggest that people are judged by their chronological age and in conformity with negative views on aging, older individuals are not expected to be active and engaged in social roles.

Other researchers were concerned with the evolution of views on aging among the German population. Studies regarding the development of views on aging over the years show that older Germans have more positive views on aging in 2008 as opposed to 1996 (Wurm and Huxhold 2012). One of the possible explanations could be the introduction of active aging policies with a focus on changing views on aging. Nevertheless, since this is a quantitative study, it provides little insight into the quality of views on aging, how did these views on aging change over time. How were views on aging shaped by the social context? In the next chapters, I will discuss tentative answers to these questions.

Critique of Active Aging Policy: Time for Formulating New Views on Aging?

The active aging policy was criticized in similar ways as the successful aging model with which it shares the same ideological base. Critiques focus on the failure of active aging policies to combat discrimination of the elderly. Some authors go as far as to claim that active aging policies sustain ageism since not all people will meet the ambitious standards of health and activity in their late life (Minkler and Holstein 2008; Martinson and Berridge 2015). Active aging norms may become coercive and people can experience negative emotions if they fail to comply to the active aging standards. The ***views on aging*** promoted by active aging policies represent dynamic, active, and engaged aging adults. Such views on aging portray older adults who practice sports, go to cultural events, participate in associations or get involved in voluntary work for the community. Failure to conform to such images of aging could mean that aging individuals are not responsible enough to care what happens to them in old age. Persons who disengage from social life become representative for the negative image of old age. This shows that positive views on aging can also indirectly be used to discriminate older individuals who for one reason or another cannot fit the profile of the active ager. In this sense, the active aging policy has been criticized for potentially marginalizing passive older individuals (Lassen and Moreira 2014; Minkler and Holstein 2008; Boudiny 2012; Boudiny and Mortelmans 2011).

Furthermore, the active aging policy was criticized for promoting the so-called individualization of risk. Namely old individuals are encouraged to take responsibility for their old age. While independence in old age is good, making older people responsible for their success or failure at aging opens the possibility for blame and discrimination. Political critiques point out that successful aging ideas can be used to facilitate cuts in welfare spending for older adults (Rudman 2006; Schultz 2001). Views on aging represent a tool through which active aging policy can promote successful aging ideas, but indirectly foster discrimination. By showing active and productive older adults as representing the ideal, aging individuals are encouraged to set ambitious goals that they may not be able to achieve.

Another critique of active aging policy refers to active aging views being defined by policy makers and not by the target population (similar to the missing voices critique of the successful aging model as described in Chap. 2). Since active aging policy seeks to create positive views on aging in a top-down manner, it does little to explore the views on aging of the population unless it has the purpose to test the effectiveness of policies. In order that the active aging strategies do not become "coercive" (Walker 2009), it was recommended to take into account the aging reality of individuals (Stenner et al. 2011). Aging views should be congruent with the everyday experiences of aging individuals. A need for older people to be directly involved in determining the role of active aging in their lives was outlined (Walker and Foster 2013).

Nevertheless, despite its criticisms, active aging policy should have a positive effect on individual aging, as well as help reduce poverty, improve health and well-being of the elderly. Introducing novel retirement or health care reforms requires changes in how one thinks about aging. All things considered, policy is closely linked with views on aging since it made use of these to promote novel ideas about old age and old persons. Thus, one can say that views on aging do play a role in policy development and implementation. Social and cultural events influence what kind of policy need to be devised. Policy uses views on aging to influence public opinion and promote novel behavioral patterns (e.g., working in ones' seventies, going to regular medical check-ups to care for ones' health, making a retirement plan). What it means to be an "old person" or to be "aging" is shaped by policy discourses emphasizing autonomy and activity. Nevertheless, the experience of aging depends on the social location and on material circumstances (Breheny and Stephens 2010; Kohon and Carder 2014) not just on culture in general (Fung 2013). The context of aging determines the barriers that people face when trying to be active and influences their health. Breheny and Stephens (2010) show that not all old individuals can adhere to the active aging ideals since their financial situation constitutes a barrier. The next chapter will examine the role of context for aging in general and for views on aging in particular. In a social reality of precarity, how will views on aging "look" like and will they foster policy change?

References

Adler, R. P., & Goggin, J. (2005). What do we mean by "civic engagement"? *Journal of Transformative Education, 3,* 236–253. https://doi.org/10.1177/1541344605276792.

Andrews, M. (1999). The seductiveness of agelessness. *Ageing & Society, 19*(3), 301–318. https://doi.org/10.1017/s0144686x99007369.

Baltes, P. B., & Baltes, M. M. (Eds.). (1990). *Successful aging.* Cambridge University Press. https://doi.org/10.1017/cbo9780511665684.

Bass, S. A., & Caro, F. G. (2001). Productive aging: A conceptual framework. In N. Morrow-Howell, J. Hinterlong, & M. Sherraden (Eds.), *Productive aging: Concepts and challenges* (pp. 37–80). Baltimore: Johns Hopkins University Press.

Barusch, A. S. (2013). The Aging Tsunami: Time for a New Metaphor? *Journal of Gerontological Social Work, 56*(3), 181–184. https://doi.org/10.1080/01634372.2013.787348.

Beck, U. (1992). *Risk society: Towards a new modernity* (M. Ritter, Trans.). London: Sage.

Boudiny, K., & Mortelmans, D. (2011). A critical perspective: Towards a broader understanding of 'active ageing'. *Electronic Journal of Applied Psychology, 7,* 8–14.

Boudiny, K. (2012). "Active ageing": from empty rhetoric to effective policy tool. *Ageing & Society, 33*(06), 1077–1098. https://doi.org/10.1017/s0144686x1200030x.

Bowling, A. (2006). Successful ageing from older people's perspectives. Results from a British survey of ageing. *European Journal of Ageing, 3,* 123–136.

Bowling, A. (2008). Enhancing later life: How older people perceive active ageing? *Aging and Mental Health, 12*(3), 293–301.

Bowling, A. (2009). Perceptions of active ageing in Britain: Divergences between minority ethnic and whole population samples. *Age and Ageing, 38,* 703–710. https://doi.org/10.1093/ageing/afp175.

Bowling, A., & Gabriel, Z. (2004). An integrational model of quality of life in older age. A comparison of analytic and lay models of quality of life. *Social Indicators Research, 69,* 1–36.

Breheny, M., & Stephens, C. (2010). Ageing in a material world. *New Zealand Journal of Psychology, 39*(2), 41–48.

Butler, R. (1975). *Why survive? Being old in America.* San Francisco, CA: Harper and Row.

Carstensen, L. L. (1992). Social and emotional patterns in adulthood: Support for socioemotional selectivity theory. *Psychology and Aging, 7*(3), 331–338. https://doi.org/10.1037/0882-7974.7.3.331.

Carstensen, L. L., Fung, H. H., & Charles, S. T. (2003). *Motivation and Emotion, 27*(2), 103–123. https://doi.org/10.1023/a:1024569803230.

Clarke, A., & Warren, L. (2007). Hopes, fears and expectations about the future: What do older people's stories tell us about active ageing? *Ageing & Society, 27*(4), 465–488.

Corsi, M., & Samek, L. (2010). *Active ageing and gender equality policies (EGGSI report for the European Commission, DG Employment, Social Affairs and Equal Opportunities).* Luxembourg: Publications Office of the European Union.

Craciun, C. (2011). Successful aging-utopia or the result of lifelong learning? Meaning and representations of ageing in Romanian elderly. *Ageing International, 37*(4), 373–385. https://doi.org/10.1007/s12126-011-9131-2.

Craciun, C., & Flick, U. (2014). "I will never be the granny with rosy cheeks": Perceptions of aging in precarious and financially secure middle-aged Germans. *Journal of Aging Studies, 29,* 78–87. https://doi.org/10.1016/j.jaging.2014.01.003.

Craciun, C., & Flick, U. (2015). "I want to be 100 years old, but I smoke too much": Exploring the gap between positive aging goals and reported preparatory actions in different social circumstances. *Journal of Aging Studies, 35,* 49–54. https://doi.org/10.1016/j.jaging.2015.07.006.

Dannefer, D. (2012). Enriching the tapestry: expanding the scope of life course concepts. *The Journals of Gerontology, Series B: Psychological Sciences and Social Sciences, 67*(2), 221–225. https://doi.org/10.1093/geronb/gbr148.

Davey, J. (2002). Active ageing and education in mid and later life. *Ageing & Society, 22,* 95–113. https://doi.org/10.1017/S0144686X02008528.

Deeming, C. (2009). 'Active ageing' in practice: A case study in East London, UK. *Policy and Politics, 37,* 93–111. https://doi.org/10.1332/030557309X397946.

Dillaway, H. E., & Byrnes, M. (2009). Reconsidering Successful Aging. *Journal of Applied Gerontology, 28*(6), 702–722. https://doi.org/10.1177/0733464809333882.

Eliassen, A. H. (2015). Power relations and health care communication in older adulthood: Educating recipients and providers. *The Gerontologist, 56*(6), 990–996. https://doi.org/10.1093/geront/gnv095.

Estes, C. L. (Ed.). (2001). *Social policy and aging: A critical perspective.* Thousand Oaks, CA: Sage.

Estes, C. L., & Phillipson, C. (2002). The globalization of capital, the welfare state, and old age policy. *International Journal of Health Services, 32*(2), 279–297. https://doi.org/10.2190/5ty7-pd68-qld1-v4ty.

Fernandez-Ballesteros, R., Olmos, R., Santacreu, M., Bustillos, A., & Molina, M. A. (2017). The role of perceived discrimination on active aging. *Archives of Gerontology and Geriatrics, 71,* 14–20. https://doi.org/10.1016/j.archger.2017.02.004.

Foster, L., & Walker, A. (2014). Active and successful aging: A European policy perspective. *The Gerontologist, 55*(1), 83–90. https://doi.org/10.1093/geront/gnu028.

Fung, H. H. (2013). Aging in Culture. *The Gerontologist, 53*(3), 369–377. https://doi.org/10.1093/geront/gnt024.

Fung, H. H., & Carstensen, L. (2006). Goals change when life's fragility is primed: Lessons learned from older adults, the September 11 Attacks and Sars. *Social Cognition, 24*(3), 248–278.

Hilton, J. M., Gonzalez, C. A., Saleh, M., Maitoza, R., & Anngela- Cole, L. (2012). Perceptions of successful aging among older Latinos, in cross-cultural context. *Journal of Cross-Cultural Gerontology, 27,* 183–199. https://doi.org/10.1007/s10823-012-9171-4.

Hui-Chuan, H. (2007). Exploring Elderly People's perspectives on successful ageing in Taiwan. *Ageing & Society, 27*(1), 87–102.

Hamblin, K. (2010). Changes to policies for work and retirement in EU15 nations (1995–2005): An exploration of policy packages for the 50-plus cohort. *International Journal of Ageing and Later Life, 5,* 13–43. https://doi.org/10.3384/ijal.1652-8670.105113.

Horton, S., Baker, J., & Deakin, J. M. (2007). Stereotypes of aging: Their effects on the health of seniors in North American Society. *Educational Gerontology, 33*(12), 1021–1035. https://doi.org/10.1080/03601270701700235.

Kohli, M. (1985). The institutionalization of the life course: Historical facts and theoretical arguments. *Kolner Zeitschrift fur Sociologie und Socialpsychologie, 37,* 1–29.

Kohli, M. (2007). The institutionalization of the life course: Looking back to look ahead. *Research in Human Development, 4,* 253–271.

Kohon, J., & Carder, P. (2014). Exploring identity and aging: Auto-photography and narratives of low income older adults. *Journal of Aging Studies, 30,* 47–55. https://doi.org/10.1016/j.jaging.2014.02.006.

Konrath, S., Fuhrel-Forbis, A., Lou, A., & Brown, S. (2012). Motives for volunteering are associated with mortality risk in older adults. *Health Psychology, 31*(1), 87–96. https://doi.org/10.1037/a0025226.

Kotter-Grühn, D. (2015). Changing negative views of aging: Implications for intervention and translational research. *Annual Review of Gerontology and Geriatrics, 35*(1), 167–186. https://doi.org/10.1891/0198-8794.35.167.

Lassen, A. J., & Moreira, T. (2014). Unmaking old age: Political and cognitive formats of active ageing. *Journal of Aging Studies, 30,* 33–46. https://doi.org/10.1016/j.jaging.2014.03.004.

Levy, B. R. (2003). Mind matters: Cognitive and physical effects of aging self-stereotypes. *The Journals of Gerontology: Series B, 58*(4), P203–P211. https://doi.org/10.1093/geronb/58.4.p203.

Levy, B. (2009). Stereotype Embodiment. *Current Directions in Psychological Science, 18*(6), 332–336. https://doi.org/10.1111/j.1467-8721.2009.01662.x.

Löckenhoff, C. E., De Fruyt, F., Terracciano, A., McCrae, R. R., De Bolle, M., Costa, P. T. Jr.,... Yik, M. (2009). Perceptions of aging across 26 cultures and their culture-level associates. *Psychology and Aging, 24,* 941–954. https://doi.org/10.1037/a0016901.

Loretto, W., & Vickerstaff, S. (2015). Gender, age and flexible working in later life. *Work, Employment & Society, 29*(2), 233–249. https://doi.org/10.1177/0950017014545267.

Mahne, K., Wolff, J. K., Simonson, J., & Tesch-Römer, C. (2017). Altern im Wandel: Zwei Jahrzehnte Deutscher Alterssurvey [Changes in Aging: Two Decades German Survey of Aging]. In K. Mahne et al. (Eds.) *Altern Im Wandel,* 11–28. https://doi.org/10.1007/978-3-658-12502-8_1.

Martinson, M., & Minkler, M. (2006). Civic engagement and older adults: A critical perspective. *The Gerontologist, 46,* 318–324.

Martinson, M., & Berridge, C. (2015). Successful aging and its discontents: A systematic review of the social gerontology literature. *The Gerontologist, 55*(1), 58–69. https://doi.org/10.1093/geront/gnu037.

Minkler, M., & Holstein, M. B. (2008). From civil rights to civic engagement? Concerns of two older critical gerontologists about a "new social movement" and what it portends. *Journal of Aging Studies, 22*(2), 196–204. https://doi.org/10.1016/j.jaging.2007.12.003.

Morrow-Howell, N. (2011). Outcomes of civic engagement for older adults. In G. O'Neill & S. F. Wilson (Eds.), *Civic engagement in an older America* (pp. 125–131). Washington, DC: Gerontological Society of America.

Mullan, P. (2004). *The imaginary time bomb: Why an ageing population is not a social problem*. London: I.B.Tauris.

Nosraty, L., Jylhä, M., Raittila, T., & Lumme-Sandt, K. (2015). Perceptions by the oldest old of successful aging, Vitality 90 + Study. *Journal of Aging Studies, 32,* 50–58. https://doi.org/10.1016/j.jaging.2015.01.002.

Naegele, G. (2016). Die Gestaltung der Spätphase des Erwerbsverlaufs – Herausforderungen für eine soziale Lebenslaufpolitik [The organization of the late phase of working life-Challenges for social policy]. In F. Freirchs (Ed.) *Altern in Der Erwerbsarbeit* [Aging and Work], 25–48. https://doi.org/10.1007/978-3-658-12384-0_2.

Naegele, G., & Walker, A. (2009). Social protection: Incomes, poverty and the reform of pension systems. *Ageing in Society: European Perspectives on Gerontology*, 142–166. https://doi.org/10.4135/9781446278918.n7.

Ouwehand, C., de Riider, D. T. D., & Bensing, J. M. (2007). A review of successful aging models proposing proactive coping as an important additional strategy. *Clinical Psychology Review, 27*(8), 873–884.

Pavlova, M. K., & Silbereisen, R. K. (2011). Perceived level and appraisal of the growing expectations for active ageing among the young-old in Germany. *Research on Aging, 34*(1), 80–99. https://doi.org/10.1177/0164027511416371.

Phillipson, C. (2013). *Ageing*. Cambridge: Polity Press. ISBN 13: 978 0 7456 3085 4.

Phillipson, C., Shepherd, S., Robinson, M., & Vickerstaff, S. (2018). Uncertain futures: Organisational influences on the transition from work to retirement. *Social Policy and Society,* 1–16. https://doi.org/10.1017/s1474746418000180.

Potratz, W., Gross, T., & Hilbert, J. (2009). The silver economy: Purchasing power and the quest for quality of life. In G. Naegele & A. Walker (Eds.), *Social policy in ageing societies: Britain and Germany compared*. Bedford, UK: Palgrave Macmillan.

Prieler, M., Kohlbacher, F., Hagiwara, S., & Arima, A. (2014). The representation of older people in television advertisements and social change: The case of Japan. *Ageing & Society, 35*(04), 865–887. https://doi.org/10.1017/s0144686x1400004x.

Riley, M. W., & Riley, J. W., Jr. (1994). Structural lag: Past and future. In M. W. Riley, R. L. Kahn, & A. Foner (Eds.), *Age and structural lag: The mismatch between people's lives and opportunities in work, family, and leisure*. New York, NY: John Wiley.

Rowe, J. W., & Kahn, R. I. (1998). *Successful aging*. New York: Pantheon.

Rowe, J. W., & Kahn, R. L. (1997). Successful aging. *The Gerontologist, 37,* 433–440. https://doi.org/10.1093/geront/37.4.433.

Rozanova, J. (2010). Discourse of successful aging in The Globe and Mail: Insights from critical gerontology. *Journal of Aging Studies, 24*(4), 213–222. https://doi.org/10.1016/j.jaging.2010.05.001.

Rudman, D. L. (2006). Shaping the active, autonomous and responsible modern retiree: An analysis of discursive technologies and their links with neo-liberal political rationality. *Ageing & Society, 26*(2), 181–201.

Schultz, J. H. (2001). *The economics of aging*. Westport, CT: Auburn House/Greenwood.

Singer, T., Lindenberger, U., & Baltes, P. B. (2003). Plasticity of memory for new learning in very old age: A story of major loss? *Psychology and Aging, 18,* 306–317.

Smirnova, M. H. (2012). A will to youth: The woman's anti-aging elixir. *Social Science and Medicine, 75*(7), 1236–1243. https://doi.org/10.1016/j.socscimed.2012.02.061.
Stenner, P., McFarquhar, T., & Bowling, A. (2011). Older people and "active ageing": Subjective aspects of ageing actively. *Journal of Health Psychology, 16*(3), 467–477. https://doi.org/10.1177/1359105310384298.
Swift, A. U., & Tate, R. B. (2013). Themes from older men's lay definitions of successful aging as indicators of primary and secondary control beliefs over time: The Manitoba Follow-up Study. *Journal of Aging Studies, 27*(4), 410–418.
Tesch-Römer, C., & Wurm, S. (2012). Research on Active Aging in Germany. *GeroPsych, 25*(3), 167–170. https://doi.org/10.1024/1662-9647/a000067.
Tesch-Römer, C. (2012). *Active aging and quality of life in old age*. New York and Geneva: United Nations.
Torres, S. (2006). Different ways of understanding the construct of successful aging: Iranian immigrants speak about what aging well means to them. *Journal of Cross-Cultural Gerontology, 21*(1–2), 1–23. https://doi.org/10.1007/s10823-006-9017-z.
Torres, S. (2014). Expanding the gerontological imagination on ethnicity: Conceptual and theoretical perspectives. *Ageing & Society, 35*(05), 935–960. https://doi.org/10.1017/s0144686x14001330.
Van den Heuvel, W. J. (2012). Discrimination against older people. *Reviews in Clinical Gerontology, 22*(04), 293–300. https://doi.org/10.1017/s095925981200010x.
Villar, F. (2012). Successful ageing and development: The contribution of generativity in older age. *Ageing & Society, 32,* 1087–1105. https://doi.org/10.1017/S0144686X11000973.
Vickers, K. (2007). Aging and the Media: Yesterday, Today, and Tomorrow. *Californian Journal of Health Promotion, 5*(3), 100–105.
Walter U., Flick U., Neuber A., Fischer C., & Schwartz F.W. (2006). *Alt und Gesund? Altersbilder und Präventionskonzepte in der ärztlichen und pflegerischen Praxis*. [Old and healthy? Views on aging and prevention in health and care practice]. VS Verlag für Sozialwissenschaften.
Walker, A. (2006). Aging and politics: An international perspective. In R. Binstock & L. George (Eds.), *Handbook of aging and the social sciences* (pp. 339–359). Burlington, MA: Elsevier Academic Press.
Walker, A. (2009). Commentary: The emergence and application of active aging in Europe. *Journal of Aging and Social Policy, 21*(1), 75–93. https://doi.org/10.1080/08959420802529986.
Walker, A., & Maltby, T. (2012). Active ageing: A strategic policy solution to demographic ageing in the European Union. *International Journal of Social Welfare, 17,* 117–130. https://doi.org/10.1111/j.1468-2397.2012.00871.x.
Walker, A., & Foster, L. (2013). Active ageing: Rhetoric, theory and practice. In R. Ervik & T. Skogedal Lindén (Eds.), *The making of aging policy: Theory and practice in Europe* (pp. 27–53). Cheltenham: Edward Elgar.
Weintritt, O. (2012). Altersbilder im Islam und unter Muslimen in Deutschland und Körperbilder im Islam. [Views on aging in Islam and among Muslims in Germany and body images in Islam] In F. Berner et al. (Eds.), *Individuelle Und Kulturelle Altersbilder* [Individual and cultural views on aging], 231–287. https://doi.org/10.1007/978-3-531-93286-6_6.
WHO. (2002). *Active ageing: A policy framework*. Geneva: Author.
Wolff, J. K., Warner, L. M., Ziegelmann, J. P., & Wurm, S. (2014). What do targeting positive views on ageing add to a physical activity intervention in older adults? Results from a randomised controlled trial. *Psychology and Health, 29,* 915–932. https://doi.org/10.1080/08870446.2014.896464.
Wolff, J. K., Beyer, A.-K., Wurm, S., Nowossadeck, S., & Wiest, M. (2018). Regional impact of population aging on changes in individual self-perceptions of aging: Findings from the German ageing survey. *The Gerontologist, 58,* 47–56. https://doi.org/10.1093/geront/gnx127.
Wurm, S., Tesch-Römer, C., & Tomasik, M. J. (2007). Longitudinal findings on aging-related cognitions, control beliefs, and health in later life. *The Journals of Gerontology, Series B: Psychological Sciences and Social Sciences, 62,* 156–164. https://doi.org/10.1093/geronb/62.3.P156.

Wurm, S., & Huxhold, O. (2012). Sozialer Wandel und individuelle Entwicklung von Altersbildern [Social change and individual devolvement of views on aging]. In F. Berner et al. (Eds.) *Individuelle und Kulturelle Altersbilder* [Individual and cultural views on aging], 27–69. https://doi.org/10.1007/978-3-531-93286-6_2.

Wurm, S., Tomasik, M. J., & Tesch-Römer, C. (2010). On the importance of a positive view on ageing for physical exercise among middle-aged and older adults: Cross-sectional and longitudinal findings. *Psychology and Health, 25*(1), 25–42. https://doi.org/10.1080/08870440802311314.

Zaidi, A., & Zolyomi, E. (2011). *'Active Ageing' Research Note 7/2011*. Brussels: Produced by the Social Situation Observatory, for the European Commission, Directorate-General for Employment, Social Affairs and Inclusion.

Chapter 4
The Social Reality of Aging

Introduction

Policy aims to shape social reality to provide its citizens with better lives. As discussed in the first three chapters, positive aging ideas outline an ideal for "a happy old age" and active aging policy attempts to make it happen. Nevertheless, the social reality may provide several challenges toward reaching positive aging goals. Discrepancies between positive aging ideals and the reality of aging experiences may shape individual and social views on aging more than policies or theoretical ideas. Thus, it is relevant to examine the social reality of aging and potential incongruities with the goals of positive aging policies.

Societies nowadays face several challenges that could hinder a happy old age such as an increasing older workforce that needs to be employed, receding pension schemes, an increasing burden on the health system or growing housing expenses. Rather than providing a positive, secure climate for growing old, all these factors imply risk and vulnerability. Furthermore, despite a political effort to promote health, activity, or independence in old age and create the ideal conditions for successful aging, several studies show an increase in inequalities among older people (Grenier and Phillipson 2018; Grenier et al. 2017; Biggs 2014; Ferraro and Shippee 2009; Price 2006). Therefore, it comes as no surprise that the present political and social context of aging is being described as a culture of risk (Grenier et al. 2017). This "new culture of risk" (Grenier et al. 2017) is best expressed by the term of "precarity". Definitions of "precarity" range from more general expressions such as "being dependent on chance and circumstances" to specific experiences such as being "at risk of falling, collapse" (Grenier et al. 2017). Regardless of the definition, being precarious or the condition of precarity always implies a state of risk, vulnerability, insecurity, or uncertainty. Millar (2017) proposed a definition classification, namely precarity as a labor category (Bourdieu 1998; Castel 2003; Kalleberg 2009; Ross 2009; Vosko 2010, 2006), as a class category (Standing 2011, Waite 2009), or an ontological experience (Butler 2004, 2009). The latter refers to precarity as a generalized condition of human life (Butler 2004). Millar (2017) argues that some studies inspired

I. C. Crăciun, *Positive Aging and Precarity*, International Perspectives on Aging 21, https://doi.org/10.1007/978-3-030-14255-1_4

by Butler's work attempted to analyze and connect the understandings of precarity both as a labor condition and an ontological experience. Namely researchers were interested in how precarious labor conditions create emotions, subjective, and lived experiences (Millar 2017). For instance, some explored how unemployment erodes social belonging (Muehlebach 2011) or determines anxiety (Molé 2010) and how youth joblessness upsets daily temporalities and life plans (Allison 2012). Following this line of thought, in my study I also considered precarity both a socioeconomic condition and an ontological experience and explored it in the context of aging.

The notion of "precariousness" is often employed in the literature to portray the consequences of liberalization policies and retreat of the welfare state (Standing 2011). However, nowadays "precarity" is considered a more appropriate term because it points out differences in experience that are socially constructed (Portacolone et al. 2018). Precarity implies a sense of insecurity following the disappearance of resources necessary for everyday life (Portacolone 2013). Portacolone et al. 2018 mentions that precarity "*evokes an intrinsic sense of insecurity resulting from coping with cumulative pressures while trying to preserve a sense of independence.*" Existential precarity entails the ephemeral nature of human lives, always at risk because of natural or man-made disasters (Butler 2004). Resources in this context refer not only to financial means but also social capital, health, or cognitive resources.

Precarity contributes to the creation of a "risky society" and a "political economy of insecurity" (Beck 1992). Since precarity has consequences on political, social, economic, and cultural levels (Millar 2017; Waite 2009; Beck and Beck-Gernsheim 2002; Bourdieu 1999), it adds to the factors that shape the experiences of old age as well as to the corresponding views on aging. Both precarity and old age are constructed in terms of risk and vulnerability. Therefore, several links can be identified between the concept of aging and the notion of precarity. According to Grenier et al. (2017), aging and old age are "*accompanied (and produced) by material (i.e. loss of income through retirement) and corporeal realities (i.e., illness, loss, the need for care) that occur at the intersection of policies, structures, discourses, and experiences of health and aging*" (see Grenier 2012 cited in Grenier et al. 2017). Although current aging discourses focus on success, activity, and health (Lassen and Moreira 2014; Stenner et al. 2011; Walker 2009), changes in the labor market and the retreat of the welfare state have made late life more insecure (Grenier et al. 2017; Phillipson et al. 2018). Thus, a discrepancy emerges between the social reality of aging and positive aging theory and policy. Precarity suggests a rather negative view on aging, where old people are seen as vulnerable, ill, and passive. In contrast, active aging policy constructs older individuals as being happy, healthy, and active. Therefore, aging individuals need to negotiate the discrepancies between negative and positive views on aging in order to find their own ways of coping both with the precarity of daily life and the insecurities of the aging experience itself.

In the current chapter, I explore the inconsistencies between the social reality of aging and positive aging discourses by analyzing views on aging as a potential link between the two. I chose to focus on three issues where discrepancies may appear and shape views on aging, namely the problem of work and retirement, health, and social integration. In this context, I examine how positive ideals of activity, autonomy, and social engagement are challenged by social realities of precarious aging. Last but not least, I go on to explore how precarity may shape cognitions and emotions, which makes the background to one of the main research questions of the present book.

Precarious Work and Vanishing Pensions

The period beginning in the 1970s was marked by uncertainty and instability on the job market that eventually led to the establishment of more insecure employment forms (Kalleberg 2013). The notion of "precarious work" was introduced to label flexible forms of employment such as contract work, part-time work, temporary employment (Kim et al. 2012). Benach, and Muntaner (2007) further described four dimensions of the relation between precarity and work including: continuity (e.g., temporary work contracts), vulnerability (e.g., powerlessness, risk for health problems), protection (e.g., limited benefits such as pension or health insurance; lack of paid holidays), and income (e.g., low-paid jobs or insecure earnings).

In an effort to find a working definition, Arne Kalleberg (2009) described precarious work as "*employment that is uncertain, unpredictable, and risky from the point of view of the worker*" (Kalleberg 2009). However, according to this characterization, in capitalist work structures, many citizens could identify their work as precarious. Thus, more objective standards were needed to delimitate precarious work from other more secure forms of employment. Several objective indicators were proposed such as the phenomenon of shifting the responsibility for work from the employers to the employee and imposing a flexible schedule or part-time employment. In the Western, developed countries, precarious jobs have a set termination date and provide no benefits such as pension or health insurance (Portacolone 2013). Thus, in addition to determining rising social inequalities, precarious work may contribute to differences in well-being between precarious and fully employed individuals (Standing 2011; Craciun and Flick 2014, 2016).

In what concerns the effects of precarious work forms on aging processes, making employees responsible means that these must take charge of their health insurance, their medical check-ups, their preventive behaviors, or their pensions. Precarious work is represented through the image of the "precarious worker." For instance, in Japan, precarious individuals are described as *the "working poor"* or *furita*, namely **irregular workers** with no job security, no benefits (e.g., pension, health insurance, paid holidays), and small wages (Allison 2013). According to Allison, the author of "Precarious Japan," the precarity of work conditions suggests a flexibilization or liquification of employment and of life in general (Allison 2013, p. 8). Such a "*precarization of life*" (Allison 2013) may have an important impact on aging experiences and expectations toward old age.

The active aging policy requires older individuals to work and "give back" to society. Nevertheless, this is not an easy task for people who do not have a stable job and lack a secure pension (Craciun and Flick 2014, 2015). On the European job market, more and more people are forced to be flexible and take part-time jobs, poorly paid jobs, or work on time limited contracts. Because of the aging of the population and crash of market economies (van Dalen et al. 2014), individuals cannot rely just on state-funded pensions anymore. Thus, middle-aged people are encouraged to find private pensions solutions instead. However, this is still a novel trend and individuals need to adapt to this new reality of precarious working lives and vanishing pensions.

During the 1980s, European seniors were urged to retire early to give a chance to younger generations and thus reduce youth unemployment (Phillipson et al. 2018). Learning, working, and resting were still regarded as three successive phases of the life course (Townsend et al. 2006) and were comprised in so called lockstep retirement formulas (Loretto and Vickerstaff 2015). Nowadays, the situation has changed, and the elderly are expected to stay employed as long as possible and remain active and engaged through continuous learning activities or volunteering (Lassen and Moreira 2014; Foster and Walker 2014). However, for individuals working in precarious jobs, it may be difficult to stay healthy and engaged in their old age. The prospect that they will have to work beyond retirement age since pension will either not be available or will not cover living expenses is rather a threatening perspective than something to look forward to. For precarious workers, finding jobs in old age constitutes a relevant issue. The question remains open whether this situation will make them happier or just more stressed in their old age. The insecurity of their pension situation can add up to whatever vulnerabilities they associate with old age. In this precarious context, the urge to be active and engaged in old age could be perceived as a stressor rather than as a path to "positive aging."

In connection with the precarization of employment forms and of pensions, negative views on aging are indirectly encouraged. Instead of positive views of "active, useful older individuals," rather negative images of older, "unemployed," "passive" persons emerge. Precarity highlights the negative aspects of working in later life by emphasizing all the risks involved (e.g., searching for work, competing with younger colleagues for jobs, pension cutbacks, not being or not feeling useful for society, not being able to plan the future). Thus, aging individuals need to handle the inconsistencies between the active aging ideal and their everyday reality of searching for new jobs or being scared to lose the one they have. In this respect, perspectives for their old age may involve precarious working scenarios or poverty in old age instead of images of blissful relaxing days spent as a retiree. Views on aging of precarious individuals may illustrate the perceived discrepancies between positive aging ideals and reality.

Precarity and (ill) Health

Depending on the definition of precarity, there are several ways in which this may influence people's health. If precarity is defined as an ontological experience (Butler 2004) where one becomes aware of ones' own vulnerability, this awareness may lead to experiencing stress and negative emotions (e.g., anxiety, fear, anger, depression etc.). There is a vast literature on the link between health and precarity defied as a work category. As mentioned above, precarious work is defined as a multidimensional construct including temporality, powerlessness, and lack of social benefits. For example, within the Whitehall II study (Ferrie et al. 2005), a combination of psychological factors (i.e., pessimism and heightened vigilance) and job characteristics (e.g., low job satisfaction) explained the relation between perceived job insecurity and self-rated health (Ferrie et al. 2005).

The health effects of job insecurity represent a novel research direction, since previous studies mainly dealt with the health effects of unemployment (Hutchings and Gower 1993) or of poverty (Braveman and Gottlieb 2014). Changes on the job market inspired research into the health effects of temporary and insecure work conditions. Insecure work contracts were shown to have a negative effect on physical and mental health (Benach et al. 2002; Ferrie 2002). Job insecurity was found to act as a chronic stressor on the mental and physical health of individuals. Working in a precarious job seems to have both stressful effects on the body (e.g., increased cardiovascular reactivity) and on the mind (e.g., chronic anxiety). Among the stress inducing factors of precarious jobs are the insecure work contract with a set termination date, the lack of benefits, diminished freedom concerning when to take a holiday or sickness leave, not being part of the organizations' identity because precarious workers do not participate in the relevant committees and usually have no right to vote. All in all, non-permanent workers were found to enjoy less autonomy and control at their work places (Benach et al. 2002). This may lead to experiencing low levels of control and autonomy at work. Furthermore, this situation may have long-lasting negative effects. It is unlikely that after being used to having little control over one's work and life circumstances, one will suddenly learn how to take charge during old age. In terms of views on aging, precarity engenders images of "dependent" older adults, rather than "active and autonomous" aging individuals as constructed by the active aging policy.

Precarious employees usually have poorer working circumstances that can have a serious impact on their health conditions (La Montagne 2010). Precarious work often includes low control and high job demands, an undisputable combination for experiencing stress at work. Job insecurity was associated with increased health risks in several European countries (László et al. 2010). A more recent study showed that while both job insecurity and unemployment were related to poor mental health, job insecurity was more strongly associated with somatic symptoms (Kim and von dem Knesebeck 2015). The authors hint that having a job is not necessarily better than having no job at all, if the job in question is precarious (Kim and von dem Knesebeck 2015). Job insecurity constitutes a chronic source of uncertainty and repeated exposure damages health in the long run. Thus, precarious individuals have a higher probability of being ill. In case of precarious workers, health disadvantages can accumulate well into their old age resulting in poor health in old age. As a consequence, health inequalities may emerge between precarious and financially secure workers making it more difficult for the former to enjoy a positive old age.

Furthermore, images of active, healthy, and independent older adults are in contrast with those of "stressed," older precarious workers. Once again, precarity generates negative images of aging that contradict positive standards of active aging.

Precarity and Loneliness in Old Age

Social engagement and social integration are among the targets of active aging policy (Stenner et al. 2011). In this context, living arrangements in old age represent an issue that is always on the policy agenda. The aim is to make sure that older people have homes that are affordable, age-appropriate, and allow people to be socially engaged (e.g., are within walking distance from shops, hospitals, community centers). Despite such efforts, there is increasing evidence of experienced loneliness in old age (Ferguson 2012; Meyer and Schuyler 2011; Azerado and Afonso 2016) and psychological distress associated with loneliness (Dahlberg et al. 2018).

Living alone in old age represents an increasingly common phenomenon (Portacolone et al. 2018). It has been registered in the USA (Stepler 2016), across Europe (Jamieson and Simpson, 2013) as well as Asia (Brown et al. 2002; Lei et al. 2015). Among the explanations offered for such trends are longer life spans, higher divorce rates, the mobility of the population (e.g., children living far away from their parents), and an increase in unmarried or single older individuals (Portacolone et al. 2018). Living alone in old age can represent a satisfactory form of independent living. However, it may also imply feeling lonely and vulnerable and thus portray a form of precarity in late life.

Loneliness in old age may be a consequence of declining support networks or the result of disaffiliation and damaged social bonds. Castel (2000) talks about precariousness as a characteristic of insecure work contracts that in time may lead to disaffiliation. In his view, people are not only vulnerable because of financial matters, but because their social relations have become insecure, and they start to feel vulnerable on a social and emotional level. Disaffiliation is defined as a rupture of society bonds, resulting in the fact that economic insecurity determines poverty and relationship insecurity becomes isolation (Castel 2000). Precarious workers who are stressed by their job situations may have limited resources to build or maintain social networks. The latter require time, energy, and often financial means.

People who live alone tend to be more involved in their social networks as those who are living with others (Klinenberg 2012). Nevertheless, with advancing age, individuals have less energy to seek out their social networks and living alone becomes associated with loneliness. According to Grenier and colleagues (2017), declining networks of support outline a new form of insecurity for old age.

When cognitive or other physical impairments come into play with age, the sense of precarity may increase. For example, Elena Portacolone explored the experience of older adults (2013) and older adults with cognitive impairments (2018) who were living alone in the USA. In the context of living alone, precarity evokes a sense of insecurity stemming from vanishing resources (e.g., income, health, social bonds, etc.). Portacolone (2013) applied a micro-, meso- and macroframework to examine the precarity of living alone in old age. At a microlevel, older people living alone struggled with their failing health and often shrinking income. At a mesolevel, older solo dwellers struggled with an incomprehensive social system and lack of family support structures, while at macrolevel, they faced the challenge of meeting up independence standards imposed by the active aging policies. The author points out the inability of existing social policies and programs to accommodate the needs of older adults living alone. The sense of precarity resulted from the need of the participants to prove that they can make it on their own (Portacolone 2011), when they would have actually needed support from services but could not afford or could not access them. In another study, Portacolone et al. (2018) showed that older individuals who suffered from a cognitive impairment and were living alone felt responsible for managing their cognitive deficiency. Findings also pointed out the lack of existing services to support such older solo dwellers. Such results show that older adults living alone may find it difficult to fit the active agers profile. Living alone constitutes a vulnerable situation in itself, and it adds to the general precarity of old age (e.g., decline in health, shrinking social networks, increased risk for poverty, etc.). Therefore, exploring living situations and loneliness in old age could provide further evidence concerning the inconsistency between a social reality marked by precarity and positive aging discourses.

Positive images of aging display older adults who are surrounded by family and friends and seniors who are engaged in activities for the good of the community. Nevertheless, precarity favors loneliness in old age and related negative images of "lonely old adults." Such negative images contradict positive views of "socially active and engaged" individuals. This contrast can affect individual and social views on aging.

How Precarity Shapes Our Cognitions and Emotions

Guy Standing (2011) in his book "The precariat: the new dangerous class" described the phenomenon of a "*precariatised mind*" and referred to some instances concerning how precariousness shapes how we think and behave. According to the author, short-term thinking constitutes a principal characteristic of precarity. In this context, short-term thinking represents an incapacity to think long-term and plan as a consequence of perceiving a low probability of career development and personal progress.

Precarious workers need to constantly look for new jobs and even contemplate continuously searching for work in their old age. Looking for jobs in the digitalized world demands a certain time spent online browsing and an increased presence on social media. Even personal relations (e.g., staying in touch with friends, looking for romance) require spending time on the Internet. Authors who criticize an increased use of technology in daily life, argue that digital behaviors rewire the brain, making people less capable of concentrating for longer periods of time and of memorizing information (Standing 2011). The interaction with the digital world often demands quick reactions and decisions and may erode the capacity to develop wisdom, thus putting in danger yet another potential positive development in old age. The electronic environment is known to encourage multitasking and thus affect the ability to concentrate for longer periods of time. Multitaskers are often found among the precariat because they have difficulties shutting out irrelevant or distracting information (Richtel 2010). Precarious workers in general are less capable to control their use of time, and this situation leads to an increase in stress. Taking all the above into consideration, one can predict that exposure to precarity makes it difficult to maintain the developmental mind, concentrated on continuous learning and long-term goals achievement.

Emotionally, precarious individuals are known to experience anger, anxiety, anomie, and alienation (Standing 2011, p. 19). The frustration emerges from working in low-paid jobs, seeing no perspectives for promotion or career advance and performing work that has no clear occupational identity (e.g., working at the call center). Anomie represents a feeling of passivity born from despair while anxiety expresses the chronic insecurity associated with jobs that can terminate at the smallest mistake.

Standing argues that the "*precariatised mind*" is motivated by fear (Standing 2011, p. 20). Nevertheless, anger and alienation may emerge not only from perceiving the insecurity of working contexts, but also from the pressure to be happy or think positively all the time (Ehrenreich 2009). Furthermore, precarious workers may be stressed by the burden to define themselves as "motivated job seekers," rather than unemployed persons. Flexibility and perceived insecurity are known to destroy feelings of trust and cooperation (Haidt 2006; Hauser 2006) at work and in personal relations, thus leading to alienation of precarious individuals.

In his book "*Liquid Times. Living in an Age of Uncertainty,*" Zygmunt Bauman (2007) uses the concept of "precarity" to define the late stage of capitalism with its flexible, contingent and irregular work patterns. For Baumann, precarity helps describe the passage from the "solid" to the "liquid" phase of modernity and its effects on human lives. Bauman associates precarity with the downfall of long-term thinking, planning and acting and "*weakening social structures where thinking, planning and acting could be inscribed for a long time to come*" (Bauman 2007, p.3). In his opinion, the collapse of planned thought and action leads to fragmented lives that are more "lateral" rather than "vertical" oriented. Moreover, the responsibility of dealing with constantly changing circumstances is left to the individual who must become a "free chooser" and bear the responsibility for one's own choices. Flexibility is required from individuals, namely a readiness to "*change tactics and style at short notice, to abandon commitments and loyalties without regret and to pursue opportunities according to their current availability*" (Bauman 2007, p. 4). Ideological changes such as the "you are on your own" neoliberal discourses imply individualism and personal responsibility in what concerns work, health, family, or pension (Kalleberg 2009, p. 3). These ideas resemble the positive aging discourse, where individuals need to take charge of their aging process. Successful aging ideas (Rowe and Kahn 1997) have been criticized for setting unrealistic standards, ignoring cultural differences (e.g., in some cultures collectivism is favored over individualism), promoting a "busy ethic" and consumerism (Liang and Luo 2012). Since planning as a cognitive resource is made redundant in a precarious context (Bauman 2007), other resources may play a key role and help people in their aging process.

All in all, precarity leaves its mark on people's cognitions and emotions. Studying how precarity shapes thought patterns and feelings can shed light on the relation between precarity and aging. Views on aging might constitute a link between precarity and aging and a reflection of what happens inside the so-called precariatized mind (Standing 2011). Precarity translates into a negative image of "anxious" and "depressed" older individuals that are in contradiction with the active aging ideal of happy, healthy, independent older persons. However, how do precarious aging workers perceive precarity and aging in precarious times? In an attempt to provide answers to such questions, within the next chapter I will explore the German middle-aged precariat and their views on precarity and aging.

References

Allison, A. (2012). Ordinary refugees: Social precarity and soul in 21st century Japan. *Anthropological Quarterly, 85*(2), 345–370.

Allison A. (2013). *Precarious Japan.* Duke University Press. http://dx.doi.org/10.1215/9780822377245.

Azerado, Z., & Afonso, N. (2016). Loneliness from the perspective of the elderly. *Rev. Bras. Geriatr. Gerontol., Rio de Janeiro, 19*(2):313–324.

Bauman, Z. (2007). *Liquid times: Living in an age of uncertainty*. Cambridge: Polity Press.

Beck, U. (1992). Risk society: Towards a new modernity (M. Ritter, Trans.). London: Sage.

Beck, U., & Beck-Gernsheim, E. (2002). *Individualization: Institutionalized individualism and its social and political consequences*. London: Sage.

Benach, J., & Muntaner, C. (2007). Precarious employment and health: developing a research agenda. *Journal of Epidemiology and Community Health, 61*(4), 276–277. https://doi.org/10.1136/jech.2005.045237.

Benach, J. M., Amable, C. Muntaner, & Benavides, F. G. (2002). The consequences of flexible work for health: are we looking at the right place? *Journal of Epidemiology and Community Health, 56*(6), 405–406. https://doi.org/10.1136/jech.56.6.405.

Biggs, S. (2014). Precarious ageing versus the policy of indifference: International trends and the G20. *Australasian Journal on Ageing, 33*(4), 226–228.

Bourdieu, P. (1998). *Acts of resistance: Against the tyranny of the market.* New York: The New Press.

Bourdieu, P. (1999). *The weight of the world. Social suffering in contemporary society*. Oxford: Polity Press.

Braveman, P., & Gottlieb, L. (2014). The social determinants of health: It's time to consider the causes of the causes. *Public Health Reports, 129,* 19–31.

Brown, J. W., Liang, J., Krause, N., Akiyama, H., Sugisawa, H., & Fukaya, T. (2002). Transitions in living arrangements among elders in Japan: Does health make a difference? *The Journal of Gerontology, Series B: Psychological Sciences and Social Sciences, 57,* S209–S220.

Butler, J. (2004). *Precarious life: The powers of mourning and violence*. London: Verso.

Butler, J. (2009). *Frames of war: When is life grievable?* London: Verso.

Castel, R. (2000). The roads to disaffiliation: Insecure work and vulnerable relationships. *International Journal of Urban and Regional Research, 24*(3), 519–535.

Castel, R. (2003). *From manual workers to wage laborers: Transformation of the social question.* Trans. Richard Boyd. New Brunswick, NJ: Transaction Publishers.

Craciun, C., & Flick, U. (2014). "I will never be the granny with rosy cheeks": Perceptions of aging in precarious and financially secure middle-aged Germans. *Journal of Aging Studies, 29,* 78–87. https://doi.org/10.1016/j.jaging.2014.01.003.

Craciun, C., & Flick, U. (2015). "I want to be 100 years old, but I smoke too much": Exploring the gap between positive aging goals and reported preparatory actions in different social circumstances. *Journal of Aging Studies, 35,* 49–54. https://doi.org/10.1016/j.jaging.2015.07.006.

Craciun, C., & Flick, U. (2016). Aging in precarious times: Exploring the role of gender in shaping views on aging. *Journal of Women and Aging, 28,* 530–539. https://doi.org/10.1080/08952841.2016.1223896.

Dahlberg, L., Agahi, N., & Lennartsson, C. (2018). Lonelier than ever? Loneliness of older people over two decades. *Archives of Gerontology and Geriatrics, 75,* 96–103. https://doi.org/10.1016/j.archger.2017.11.004.

Ehrenreich, B. (2009). *Smile or die: How positive thinking fooled America and the World.* London: Granta.

Ferguson, L. (2012). Tackling loneliness in older age – Why we need action by all ages. *Quality in Ageing and Older Adults, 13*(4), 264–269. https://doi.org/10.1108/14717791211286931.

Ferraro, K. F., & Shippee, T. P. (2009). Aging and cumulative inequality: How does inequality get under the skin? *The Gerontologist, 49*(3), 333–343. https://doi.org/10.1093/geront/gnp034.

Ferrie, J. E., Shipley, M. J., Newman, K., Stansfeld, S. A., & Marmot, M. (2005). Self-reported job insecurity and health in the Whitehall II study: Potential explanations of the relationship. *Social Science and Medicine, 60*(7), 1593–1602. https://doi.org/10.1016/j.socscimed.2004.08.006.

Ferrie, J. E. (2002). Effects of chronic job insecurity and change in job security on self reported health, minor psychiatric morbidity, physiological measures, and health related behaviours in British civil servants: the Whitehall II study. *Journal of Epidemiology and Community Health, 56*(6), 450–454. https://doi.org/10.1136/jech.56.6.450.

Foster, L., & Walker, A. (2014). Active and successful aging: A European policy perspective. *The Gerontologist, 55*(1), 83–90. https://doi.org/10.1093/geront/gnu028.

Grenier, A., & Phillipson, C. (2018). Precarious aging: Insecurity and risk in late life. *Hastings Center Report, 48,* S15–S18. https://doi.org/10.1002/hast.907.

Grenier, A., Phillipson, C., Laliberte Rudman, D., Hatzifilalithis, S., Kobayashi, K., & Marier, P. (2017). Precarity in late life: Understanding new forms of risk and insecurity. *Journal of Aging Studies, 43,* 9–14. https://doi.org/10.1016/j.jaging.2017.08.002.

Grenier, A. (2012). *Transitions and the lifecourse: Challenging the constructions of 'growing old'.* Bristol: Policy Press.

Haidt, J. (2006). *The Happiness Hypothesis.* London: Arrow Books.

Hauser, M. D. (2006). *Moral minds: How nature designed our universal sense of right and wrong.* New York: Harper Collins.

Hutchings, J., & Gower, K. (1993). Unemployment and mental health. *Journal of Mental Health, 2*(4), 355–358. https://doi.org/10.3109/09638239309016971.

Jamieson, L., & Simpson, R. (2013). *Living alone: Globalization, identity, and belonging.* Basingstoke: Palgrave MacMillan.

Kalleberg, A. L. (2009). Precarious work, insecure workers: Employment relations in transition. *American Sociological Review, 74*(1), 1–22. https://doi.org/10.1177/000312240907400101.

Kalleberg, A. L. (2013). Book review symposium: Response to reviews of Arne L Kalleberg, good jobs, bad jobs: The rise of polarized and precarious employment systems in the United States, 1970s to 2000s. *Work, Employment & Society, 27*(5), 896–898. https://doi.org/10.1177/0950017013480403.

Kim, I. H., Muntaner, C., Vahid Shahidi, F., Vives, A., Vanroelen, C., & Benach, J. (2012). Welfare states, flexible employment, and health: A critical review. *Health Policy, 104*(2), 99–127. https://doi.org/10.1016/j.healthpol.2011.11.002.

Kim, T. J., & von dem Knesebeck, O. (2015). Is an insecure job better for health than having no job at all? A systematic review of studies investigating the health-related risks of both job insecurity and unemployment. *BMC Public Health, 15*(1). https://doi.org/10.1186/s12889-015-2313-1.

Klinenberg, E. (2012). *Going solo*. New York: Penguin Press.

La Montagne, A. D. (2010). Commentary: Precarious employment: Adding a health inequalities perspective. *Journal of Public Health Policy, 31*(3), 312–317. https://doi.org/10.1057/jphp.2010.25.

Lassen, A. J., & Moreira, T. (2014). Unmaking old age: Political and cognitive formats of active ageing. *Journal of Aging Studies, 30,* 33–46. https://doi.org/10.1016/j.jaging.2014.03.004.

László, K. D., Pikhart, H., Kopp, M. S., Bobak, M., Pajak, A., Malyutina, S., … Marmot, M. (2010). Job insecurity and health: A study of 16 European countries. *Social Science and Medicine, 70*(6), 867–874. https://doi.org/10.1016/j.socscimed.2009.11.022.

Lei, X., Strauss, J., Tian, M., & Zhao, Y. (2015). Living arrangements of the elderly in China: Evidence from the CHARLS national baseline. *China Economic Journal, 8,* 191–214. https://doi.org/10.1080/17538963.2015.1102473.

Liang, J., & Luo, B. (2012). Toward a discourse shift in social gerontology: From successful aging to harmonious aging. *Journal of Aging Studies, 26*(3), 327–334.

Loretto, W., & Vickerstaff, S. (2015). Gender, age and flexible working in later life. *Work, Employment and Society, 29*(2), 233–249. https://doi.org/10.1177/0950017014545267.

Meyer, R. P., & Schuyler, D. (2011). Old age and loneliness. The Primary Care Companion For CNS Disorders. https://doi.org/10.4088/pcc.11f01172.

Millar, K. M. (2017). Toward a critical politics of precarity. *Sociology Compass, 11*(6), e12483. https://doi.org/10.1111/soc4.12483.

Molé, N. J. (2010). Precarious subjects: Anticipating neoliberalism in northern Italy's workplace. *American Anthropologist, 112*(1), 38–53.

Muehlebach, A. (2011). On affective labor in post-Fordist Italy. *Cultural Anthropology, 26*(1), 59–82.

Phillipson, C., Shepherd, S., Robinson, M., & Vickerstaff, S. (2018). Uncertain Futures: Organisational Influences on the Transition from Work to Retirement. *Social Policy and Society*, 1–16. https://doi.org/10.1017/s1474746418000180.

Portacolone, E. (2011). The myth of independence for older Americans living alone in the Bay Area of San Francisco: A critical reflection. *Ageing & Society, 27,* 166–174. https://doi.org/10.1016/j.jaging.2013.01.001.

Portacolone, E. (2013). The notion of precariousness among older adults living alone in the U.S. *Journal of Aging Studies, 27*(2), 166–174. https://doi.org/10.1016/j.jaging.2013.01.001.

Portacolone, E., Rubinstein, R. L., Covinsky, K. E., Halpern, J., & Johnson, J. K. (2018). The precarity of older adults living alone with cognitive impairment. *The Gerontologist*. https://doi.org/10.1093/geront/gnx193.

Price, D. (2006). Why are older women in the UK poor? *Quality in Ageing and Older Adults, 7*(2), 23–32. https://doi.org/10.1108/14717794200600011.

Richtel, M. (2010, June 7). Hooked on gadgets, and paying a mental price. *New York Times,* p. 1.

Ross, A. (2009). *Nice work if you can get it: Life and labor in precarious times*. New York: New York University Press.

Rowe, J. W., & Kahn, R. L. (1997). Successful aging. *The Gerontologist, 37,* 433–440. https://doi.org/10.1093/geront/37.4.433.

Standing, G. (2011). *The precariat: The new dangerous class*. London: Bloomsbury Academic.

Stepler, R. (2016). *Smaller share of women 65 and over are living alone*. Washington, DC: Pew Center.

Stenner, P., McFarquhar, T., & Bowling, A. (2011). Older people and "active ageing": Subjective aspects of ageing actively. *Journal of Health Psychology, 16*(3), 467–477. https://doi.org/10.1177/1359105310384298.

Townsend, J., Godfrey, M., & Denby, T. (2006). Heroines, villains and victims: Older people's perceptions of others. *Ageing and Society, 26*(6), 883–900.

Van Dalen, H. P., Henkens, K., & Wang, M. (2014). Recharging or retiring older workers? Uncovering the age-based strategies of European employers. *The Gerontologist, 55*(5), 814–824. https://doi.org/10.1093/geront/gnu048.

Vosko, L. F. (Ed.). (2006). *Precarious employment: Understanding labour market insecurity in Canada*. Montreal and Kingston: McGill-Queens University Press.

Vosko, L. F. (2010). *Managing the margins: Gender, citizenship, and the international regulation of precarious employment*. Oxford: Oxford University Press.

Waite, L. (2009). A place and space for a critical geography of precarity? *Geography Compass, 3*(1), 412–433. https://doi.org/10.1111/j.1749-8198.2008.00184.x.

Walker, A. (2009). Commentary: The emergence and application of active aging in Europe. *Journal of Aging and Social Policy, 21*(1), 75–93. https://doi.org/10.1080/08959420802529986.

Part II

Chapter 5
Perceptions of the Precarious Life: The Middle-Aged Precariat in Germany

Introducing the Precariat

Guy Standing (2011) defines the precariat as a new social class that he labels a "*class in the making*" (italics used by Guy Standing) by applying a mixture of precarité and the proletariat. It is, however, difficult to define the precariat as a distinctive social class, where one either belongs or does not belong to. Since precarious work is quite volatile, a person may move in and out of the precariat in a flexible way (which is also a characteristic of precarity itself).

Belonging to the precariat is different than being part of the middle class. Being part of the precariat means one does not have a stable job in terms of financial means and does not enjoy a certain status or employment contract. For example, a doctor who is self-employed and owns a private practice, although taking financial risks, has the financial means to afford a decent living and may invest in private pension funds. Moreover, a doctor with a private practice will enjoy a certain respect and status in the community where he or she is working. Nevertheless, for example cleaners working in the same private practice will not have the same status and respect, no matter how important their work is. Medical interns who work for little money to gain experience, or doctors who work on a short-term contract with no prospects of being hired later will not enjoy the status and security as permanently employed doctors.

Nowadays, many individuals who hold a high education diploma are forced by market economy trends to work in precarious jobs or to be underemployed. In this sense, the precariat does not possess a clear occupational identity. For example, working at a call center requires no specific vocational qualifications and has little perspectives for career development. One can work in such jobs as longs as one is young and fit. However, aging processes might make it difficult to sustain the same level of energy and motivation required for such jobs.

Certain sub-populations are considered more likely to end up being precarious. Women, young people, migrants are among the categories most frequently mentioned in the literature (Buffel and Philipson 2011; Standing 2011). For example, Standing

I. C. Crăciun, *Positive Aging and Precarity*, International Perspectives on Aging 21, https://doi.org/10.1007/978-3-030-14255-1_5

(2011) describes the young and old as well as women or migrants in general as being "precarious" or belonging to the "precariat." Thus, the precariat may not be defined by a distinctive age or education level, but it might be gendered since women are usually the ones who sacrifice work goals for family purposes. Women and migrants represent the most frequent categories of the precariat in the US and European contexts alike (Kalleberg 2013). Nevertheless, it is difficult to give the precariat a certain face, since people with different education levels, social or cultural backgrounds might end up in the "precarity trap."

Who Are the Aging Precariat and Why Are They Important?

According to the life course perspective, disadvantages accumulate over the life span and contribute to higher inequalities in old age (Ferraro and Shippee 2009; Dannefer 2003). The authors state that social systems generate inequality and this manifest itself across the life course through demographic and developmental processes. According to such theory, growing up in precarious circumstances would lead to an accumulation of disadvantages which would be even greater in late adulthood. Thus, precarious older individuals would experience greater disadvantages than younger ones. For example, unemployment creates social isolation and poverty. The latter makes it difficult for people to re-enter employment and makes them vulnerable to social exclusion (Gallie et al. 2003). Nevertheless, psychological resources and human agency play a vital role in how our lives are shaped despite being subjected to social adversity (Ferraro and Shippee 2009).

Income in later life is known to be relevant for experiencing well-being, health, and social engagement (Smith et al. 2004). However, there is evidence that the elderly are poor, especially women more so than men (Price 2006; Bardasi and Jenkins 2010; Bardasi et al. 2002). For example, in the UK, among the reasons for this situation are the fact that women do not benefit as much from the pension system as men, and women are more involved in unpaid care (e.g., taking care of children, of ill relatives or doing housework). This scenario was feasible when the majority of men enjoyed full employment and thus received full pensions that allowed them to support their wives financially in old age (Price 2006). However, with neoliberal transformations, state pensions were drastically reduced and therefore poverty increased among older individuals. It also fostered the growing privatization of the pensions system. The gender pay gap, the fact that women were being paid less for the same amount of work as compared to men, further contributes to increased poverty for women in old age (Harkness 2002). Furthermore, women are known to live on average longer lives and thus they represent the majority among the older population. Several women also live alone in old age since they are widowed, separated, or divorced (Arber and Ginn 2004). Thus, it is not surprising that more older women live in poverty as compared to older men (Price 2006).

As discussed in previous chapters, after decades in which early retirement was endorsed, the aging generation is encouraged to work for longer years (Foster and

Walker 2014). Since older people are less interested in career building, they are more likely to take on precarious jobs and thus contribute to expanding the precariat. Standing (2011) describes the aging precariat as being called either "groaners" or "grinners." The grinners just want to keep active since they have no financial need to keep working after retirement age. They have paid their mortgages, and they have a pension to fall back on. Such individuals will take on any job in old age that provides them with a sense of meaning and allows them time to enjoy their late years. For example, a retired bank manager may work as a part-time social worker in an NGO. However, by doing this, such older individuals deliberately take jobs that are way below their potential and compete with young people who also want these jobs. By contrast, the groaners have a small pension and probably some debts to the bank. Consequently, they need to take on jobs to pay their living expenses in old age. The aging precariat may constitute a threat to the precarious youth since they will take on jobs that are beneath their capabilities and knowledge just because they need them and because careers do not matter to them. Older migrants constitute another specific social group who usually ends up performing precarious work (Axelsson et al. 2017; McDowell 2013).

In the USA, non-voluntary part-time work has increased after the economic recession in 2008, especially among workers older than 55 (Reynolds and Wenger 2010). This situation led to an increased risk of financial and health problems among older individuals. In Europe, the situation is very similar. For example, in Germany, part-time employment (i.e., being employed or remunerated for less than 30 h a week) constituted a solution to control unemployment after the fiscal crisis. Since many Germans were shifted into so-called mini jobs (i.e., part-time jobs, remunerated with around 400 Euros per month), this maintained the illusion of high employability for a while (Standing 2011). In the following, I will analyze the aging German precariat in more depth.

The Aging Precariat in Germany

Researchers agree that defining "precarious employment" in the German context is a complicated endeavor (Weinkopf 2009; Bispinck and Schulten 2011; Brady and Biegert 2017). The direct translation in German would be "*prekäre Beschäftigung*" (i.e., precarious work). However, this phrase is not often utilized (Bäcker 2007; Buntenbach 2007), and "atypical employment" is used instead (Weinkopf 2009; Bispinck and Schulten 2011). Precarious jobs in Germany include regular part-time work, temporary agency employment, fixed-term contracts, and the so called "mini jobs." The latter represent a particular form of part-time employment with monthly gross earnings of around 400 Euros and no required to pay taxes and social security contributions (Bispinck and Schulten 2011; Weinkopf 2009). Some forms of self-employment (e.g., low skilled, low paid) are also considered precarious jobs and called "economically dependent self-employment" (Bispinck and Schulten 2011). For the present research, we consider the definition of Kalleberg (2009, 2011) who describes precarious employment as uncertain, insecure, lacking in social protection, and rights that are given to full-time or standard employment relationships. The main characteristics of precarious employment would include low or intermittent payment, low employment stability, an indeterminate work schedule (e.g., work shifts keep changing, work hours are actually longer than stated in the contract, etc.), limited access or even exclusion from the social security system. Since precarity seems to take so many forms, one may ask if precarious employment is a common phenomenon in Germany?

Studies show that the proportion of working poor is increasing in Germany (Brady and Biegert 2017; Broding et al. 2010; Bispinck and Schulten 2011). Since the mid-eighties, precarious employment has steadily increased and in 2011 amounted to one third of the working population (Bispinck and Schulten 2011). For example, according to the German Federal Statistical Office, the number of part-time workers virtually doubled from 4.7 million in 1991 to 9.2 million in 2010 (Statistisches Bundesamt 2011). The percent of part-time workers from the whole number of German workers also increased from 14% at the beginning of the nineties to 27% in 2009 (Statistisches Bundesamt 2011). Furthermore, there was a steady increase in temporary work contracts. For instance, in 2001, 32% of the newly hired were offered a permanent work contract, whereas in 2009, 47% of the newly hired were employed with a temporary job contract (Hohendanner 2010). The younger age group was over-represented among the temporary workers, with almost half of the workers aged 20 to 25 being employed on a fixed-term contract (Hohendanner 2010). Yet, one needs to take into consideration that these people have a high chance of staying all their working lives on fixed-term contracts and thus becoming part of the aging precariat, a group “at risk” for health problems and reduced security in old age. Precarious work occurs among employees and among self-employed workers alike (Bührmann and Pongratz 2010). The precarity of self-employment results from the working conditions (e.g., often low payment, long working hours), indeterminate, and irregular volume of work and incomplete access to security system (Kock 2008). Because of the irregularity of self-employment, it is also more difficult to have access to clear statistics concerning the number of self-employed workers.

Taking the evidence of the rise in precarious employment into consideration, it is relevant to explore how did it happen? Why do people end up working and living in precarious circumstances? All in all, the restructuring of the labor market encouraged the replacement of the full-time jobs with all included benefits with precarious employment such as part-time jobs, home-based work, self-employment, temporary work, and subcontracting (Simonson et al. 2015; Bispinck and Schulten 2011). As stated above, the restructuring of the labor market constituted a measure to reduce unemployment after the financial recession.

During the last decades, Germany has undergone profound changes and the current labor market differs from the idealized image of economic accomplishment and relative equality (Brady et al. 2010). Germany represents an interesting case study since it illustrates the development of the coordinated market economy (Brady and Biegert 2017; Unger 2015; Thelen 2012). During the 2000s economic growth improved and the unemployment rate decreased (Thompson 2013) after a period where it had reached one of the largest levels in Europe (Vitols 2004). This was labeled by some researchers the "German miracle" (Burda 2016). However, behind this miracle a much bleaker picture appears, namely one characterized by an increase in precarious jobs and rising inequality (Brady et al. 2010; Standing 2011). In relation to this issue, Brady and Biegert (2017) examined the causes of the rise in precarious employment in Germany. The authors reached the conclusion that institutional changes are more responsible for this development as compared to demographics, education, skills, work, and job characteristics (Brady and Biegert 2017). Brady and Biegert (2017) argue that such institutional factors cannot be explained only by the challenges posed by the reunification of Germany. In their view, potential explanations include rather the decline of unionization (e.g., membership, coverage) and the introduction of labor market reforms designed in the spirit of activation and aiming to reduce unemployment (e.g., the "Hartz IV"). The labor market activation reforms are similar to active aging policies who encourage activity and productivity in old age instead of "being dependent." These reforms are said to have led to a dualization of the labor market into insiders and outsiders (Biegert 2014) differentiated by compensations, benefits, and protections (Brady and Biegert 2017). The outsiders are the precarious workers. These reforms have enabled the spread of atypical employment (e.g., fix-term contracts) and mini jobs.

In terms of the present research, it is relevant to ask who is affected by precarity in Germany, and more specifically among the aging precariat? A good deal of precarious work (e.g., with irregular working hours, no health and pension benefits) is performed by immigrant workers or workers with little education (Broding et al. 2010) and women (Weinkopf 2009; Bylow and Vaillant 2014; Schaefer and Gottschall 2015). In relation to the gender bias among the precariat, the Federal Employment Agency in Germany reported in 2010 that 83% of the regular workers (i.e., financially secure) were men, while 66% of the part-time workers (i.e., precarious workers) were women (Bundesagentur für Arbeit 2011). Germans from former East Germany are more affected by changes on the labor market (Simonson et al. 2011). Social inequalities in old age were reported to be on the rise, and the poverty risk for older adults has increased. Older workers in precarious employment are known to experience more physical and psychological stress. This situation may lead to an increase in health inequalities in old age between precarious individuals and their non-precarious counterparts (e.g., full-time employees, property owners). During the last decades, an increase has been reported for the number of older Germans working post-retirement age (Naumann and Romeu Gordo 2010). Nevertheless, among the reasons provided for taking up jobs in old age were not just financial needs but also the desire to give meaning to retirement by continuing with work (Engstler and Romeu Gordo 2016). This fact was interpreted to be caused also by a change in images of aging, better health, and education level of the aging generations (Mahne et al. 2016).

With a large aging baby boomer generation approaching retirement, Germany faces the challenge of poverty in old age (Simonson et al. 2012; Geyer and Steiner 2010). In 2013, there were 13 million baby boomers who were reaching retirement age. The baby boomers represent the first generation who will be affected by policies such as the rise in pension age to 67 years old and the privatizations of pensions. Findings from the DEAS panel (German Aging Survey) point out that the baby boomers, born between 1956 and 1965, will have a worse financial situation compared to the present retirees who are born after the Second World War (Simonson 2013). This is partially explained by the work trajectories of the baby boomers that are characterized by frequent job changes, part-time contracts, occasional unemployment, and freelance jobs (Kelle et al. 2012; Simonson et al. 2011; Buchholz and Blossfeld 2009). For women in particular, re-entering the job market after childbirth was difficult and several have chosen part-time contracts to be able to dedicate themselves to childcare (Buchholz and Grunow 2006). Even if there are fewer women who are just housewives, their part-time jobs make them vulnerable or even exclude them from the job market (Buchholz and Blossfeld 2009). Consequently, many of these individuals do not have a secure pension plan and can be described as part of the aging precariat. In 2008, only 15% of the baby boomers from the study reported they have private pension insurance or other financial resources such as real estate or a sum larger than 12.500 euro. An important part of the German baby boomers is represented by the freelance workers. All in all, in Germany the freelancer population has almost doubled from 1991 to 2005 reaching 2.29 million (Ziegelmeyer 2013). Findings from the SAVE (Sparverhalten und Altersvorsorge in Deutschland 2005–2008) show that freelancers with small earnings tend to save little for their old age.

Although there are several quantitative studies related to precarious work (McNamara et al. 2011; Kalleberg 2013), there is still little qualitative evidence (Portacolone 2013, 2017; Craciun 2018) especially concerning how people with precarious jobs perceive themselves and their activities. In the present chapter, I will present findings from my study on precarious middle-aged Germans to show how they perceive their precarious employment and precarity in general. In the next chapters, I will go on to discuss the relation between precarity and views on aging as well as preparatory efforts toward a good old age.

Why Berlin and the Chosen Case Studies

Berlin is the capital city of Germany, but by far not one of the richest cities in the country. According to the European Commission, Berlin has a disposable income below the national average (EURES 2018). Furthermore, Berlin has a high percentage of people employed in precarious jobs and was called in the press by a representative from ver.di, one of Germany's biggest trade union, "the capital of the working poor" (Euronews 2017). Described by its former mayor as "poor but sexy," Berlin attracted creative personalities, students, immigrants, and "alternative" types who wanted to try their luck in the big city. Berlin is also the home for a large population of immigrants in Germany and unemployed individuals. Thus, this study location permitted to recruit diverse types of middle-aged Germans who experience both precarity and aging.

The study sample comprised both Germans from the former BDR (West Germany) and DDR (East Germany), people with migration background, persons who are highly educated and those who only had a vocational training. An equal number of women and men were interviewed, all aged between 45 and 65 years old. All participants were living in Berlin at the time of the study (2013–2014) and were employed either in precarious job or had secure financial assets for old age (e.g., real estate, secure work contracts). However, their educational and life experience backgrounds were different. For example, some were living alone, while others had families and children. Another common feature was the fact that participants were middle-aged and had reached a point in their lives when they started thinking about aging.

In the following, I will explore how the interviewed precarious middle-aged Germans perceive their life and how they joined the precariat. The advantages and disadvantages of the precarious life will be presented from the perspective of the interviewees, and the connections they see between aging and precarity will be explored. The chapter will conclude with an examination of how participants coped with their precarious life.

"Joining" the Precariat

Despite a growing literature on precarious work and its effects on health (Kim et al. 2012; Kim and von dem Knesebeck 2015; Malenfant et al. 2007), there is still little knowledge on how precarious individuals perceive their situation (e.g., how they explain the fact that they are working and living in precarious circumstances). Since the path to precarity may vary from individual to individual, I wanted to understand how the study participants become precarious workers in the first place. Moreover, I was interested to know how the interviewees perceived their work situation. Thus, I asked the interviewees to describe what they are doing professionally and how their daily routine looks like. In addition, I asked them to explain how they perceived their current employment form and how they believe it affects their aging.

When explaining how they ended up in the situation of precarity, two main paths were described by the interviewed precarious Germans. One refereed to the issue of "individual choice," emphasizing that a precarious job was consciously chosen at one point in their lives. A second path refers to the issue of determination, of being part of a "surprised generation" that ended up in precarious circumstances while approaching old age.

The Paradox of Freedom

Even though all participants worked in precarious jobs, their employment choices were quite different. In what concerns their explanations about entering the precariat, one of the main themes was that of ***individual choice***. Part-time jobs or self-employment forms were chosen as a reaction to the existing work culture and strict norms about being "*present at work*" (in German the so-called Anwesenheitskultur). In contrast to a job that offered little schedule flexibility (e.g., one needs to be at the office every day from nine to five o'clock), a freelance job initially offered many advantages like work schedule flexibility and "being one's own boss." Furthermore, for some participants, their precarious job continued to present some advantages even after years spent in precarity. Also, for several of the interviewed individuals, working in a precarious job did not automatically mean they would describe themselves as part of a particular disadvantaged social class (i.e., the precariat). Being middle-aged and still in good health made them feel that a precarious job (e.g., part-time or on a fixed contract, freelance activity) provided them with the desired freedom. For example, some participants believed they could earn enough by freelancing and still enjoy leisure time for themselves. When talking about themselves and their attitude toward work and life, participants who considered they "chose" the precarious jobs were confident they could always find solutions. The interviewees thought that their main strength is their capacity to improvise and cope with the precarious conditions of their lives. However, even if they found temporary solutions in the present, the precarity threat was perceived as real when they talked about their future. In one way or another, all participants mentioned precarity in connection to aging and talked about how it would be more difficult to find work in old age. The interviewees had no illusions of living from their pensions and wondered how long they would be able to work to earn their living. Most participants considered that even if at the moment they would not necessarily describe themselves as part of the precariat, they would surely "be precarious" in their old age.

Originally from a smaller town in West Germany, **Michael**, a self-employed teacher in his late fifties, talked about his choice of becoming a freelancer despite being aware of the disadvantages it implied. Thus, he described his choice for a freelancing employment as being a conscious one and a manifestation of his individuality. Nevertheless, in his view, being a freelancer in Germany means that one is responsible of one's own health and pension insurance. As long as one can manage to be gainfully employed, one can enjoy the freedom of the flexible schedule and self-management. However, there is always the anxiety that one may not manage to make enough money to cover all living expenses. In this context, **Michael** described this feeling of anxiety as *the paradox of freedom*. Namely in his view, being self-employed allowed him to choose when and what he worked, but he had to bear all the costs of this freedom. Not having a long-term work contract with an organization, meant freedom of choice in what concerns the work schedule and what type of work he does. Nevertheless, Michael had no guarantee for health and pension benefits and sometimes had to invest more energy and time into work, while getting less money for his efforts. However, the belief that he gets to choose what kind of work he does, kept up his motivation. Sometimes, the possibility of choice even compensated for smaller financial reimbursements:

> If you do not have a contract then of course you are not sure if you have work next month, they might not need you anymore or there is just no work. But I saw that over the years it worked somehow with a bit of luck. Of course, I work a lot more than if I had a regular work contract, the thing is one actually works more for less money. For example, for doctors

> who have their own practice it would be different. Despite these drawbacks, I believe a big advantage is the fact that you can keep the feeling you chose to go to work, you wake up in the morning and I know I must work to make money, but I chose my work, there is no employer to order me about. (…)

Michael went on to explain how the freedom "to be one's own boss" also has its shortcomings. He acknowledged the fact that having a boss to hate and criticize may be easier than having no boss at all but oneself. In his opinion, when you are your own boss you need to criticize yourself when things are not going well, and this can be very stressful. Thus, the freedom of choice becomes the burden of responsibility. Having the freedom to make all choices concerning work also comes with the responsibility for the outcomes and one has no boss to hate and blame but oneself:

> you are your own boss, but this can also be a bit unhealthy because then you pressure yourself to do stuff, and this can have negative consequences, bodily and psychologically. It is more difficult to rebel against oneself then against an external boss.

Moreover, one is not as free as one wishes to be or as one would like to believe one could be. In the end, when working in precarious circumstances, if one has to choose between doing some work and leisure time activities, one usually must choose work:

> And sometimes it is illusory this freedom, but still important. For example, if I want to go to a concert, but the language school offers me work on that day, I know I could say no but, in the end, I take the job and give up the concert.

All in all, even if future insecurity and sometimes present insecurity represent big shortcomings of freelancing jobs, the freedom of being "one's own boss" and maintaining an "illusion of freedom" represent big advantages in **Michael's** view and often outweigh potential shortcomings.

Similar to Michael, **Corinna**, a self-employed business woman in her mid-fifties, described how she became a freelancer by choice. When she first started working, **Corinna** was employed on a secure work contract as she considered to be "typical" for her generation. Nevertheless, she soon realized she would prefer to be self-employed and manage her own work schedule. She justified her choice with the ***freedom*** she gained by having a feeling of control over her time. Just like Michael, she acknowledged that most of the time her freedom of choice was rather deceptive. However, it still represented an illusion that she deeply cherished:

> I could not work every day in an office, for the same organization, every day from 9 to 18. I like to decide when to take a break from work, when to start work in the morning, I want to decide when to end my workday in the evenings. Of course, this autonomy can sometimes be an illusion. Still, I am very much bothered by the German attendance ethos where you are evaluated by the quantity of hours you spend at work and not by the quality of the products you deliver. So, the freedom of organizing your time is important for me and the fact that I can chose which organisations I want to work for and do fundraising activities. I have a huge need for independence and freedom. So, for me it is difficult to work for just one single organization.

Corinna believes her life choices came as a natural and conscious rebellion against her parents' generation. In her family in a small West German town, she was exposed to classical work patterns and traditional gender role models. Her father worked all his life and retired with 58 and her mother was a housewife. Her mother "retired" from her house chores when all her four children left home.

Corinna considered herself atypical for her generation. Most of her former class mates remained in their home town and kept the same job all their lives. In her opinion, many of her school class mates were too accustomed to the atmosphere of security and stability that they grew up with and that is why they chose to remain in the same jobs and in their hometowns their whole life. In **Corinna's** view, leaving the safety net and looking for ones' own path requires a certain courage. The same pattern of "rebellion" against the mainstream trend emerged in case of other precarious middle-aged German interviewees. Participants with precarious jobs saw themselves as part of a generation who could choose between stable employment and freelancing (i.e., precarious work). For them, making their own choices constituted an expression of freedom.

Like Michael, **Corinna** mentioned the paradox of freedom that comes with precarious employment. She described her everyday activities as a dynamic and constant quest for new clients and projects. The motto for her day is "*stay active and find work projects.*" The work flow never stops. Every evening she goes to several events where she can find potential clients for her fundraising work and where she might have a better chance to be hired on a part-time contract. Since most events where she can find work take place on Saturday and Sunday, she usually works more on weekends than during the week. For her, the concept of weekend does not really exist. Theoretically she can take time off whenever she likes (e.g., she could take some free days during the week and call it a "weekend"). However, **Corinna** only feels entitled to a free day when she feels "she deserves it." Consequently, most of the time she ends up not taking days off. Like in Michael's case, freedom for Corinna implies also the responsibility of making the "right choices" and "taking the blame" when something goes wrong with her work. Nevertheless, as in Michael's case some of the negative aspects of being self-employed only became apparent in time and **Corinna** had not foreseen them when she came to Berlin and first started freelancing.

Apart from the responsibility implied by freelance jobs, the financial shortcomings can also be stressful. For example, **Corinna** mentioned how her psychological reactions to financial insecurity surprised her:

> In my first years of self-employment all went smoothly, but then I had a period when I had no work and was penniless. At that time, I realised what it does to my self-esteem, it is difficult to remain confident when you know that your work is not required. I find this feeling difficult that I must wait for better times to come and then nothing good happens. I talked to colleagues and they also do not understand the current economic situation. I find this uncertainty burdensome. I am also in my late forties, so I must start thinking about old age and what I will survive on in my late years. It was for me a very abstract topic, but last year I was very ill, so I realised I will not always have the energy I have now. That is when I started to worry if I will be able to work in my old age.

What **Corinna** describes lends support to previous studies that showed intermittent employment contributes to reduced self-esteem of workers, loss of perceived social roles, and even an increase in depressive symptoms or suicidal ideation (Malenfant et al. 2007; Kim et al. 2012; Han et al. 2017; Min et al. 2015). For instance, a qualitative study from Canada showed that intermittent workers cannot develop a work identity, and this often demotivates them and leads to their detachment from paid labor. Others accept compromises to continue working, but the work they do often contradicts their values and thus contributes to their low self-esteem and their overall low well-being (Malenfant et al. 2007).

The cases described above are all middle-aged Germans that came from the former West German Republic. However, among precarious workers one can find also several middle-aged individuals that originally came from East Germany. **Alexandra** is a self-employed interpreter in her early sixties. Her story resembles those of Corinna and Michael. The difference would be that she is originally from East Germany and started working in West Germany during the nineties. The early nineties were a time for new beginnings and opportunities for East Germans as well as for West Germans. As a "wind of change was blowing," Alexandra made the decision to go to West Germany to find a job. At first, she worked in regular, stable employment, but then certain circumstances made her realize she needed more freedom. When talking about her life choices, **Alexandra** just as Corinna and Michael, noticed at some point in her work life, that she was not made for regular employment. Certain life events also contributed to her work decisions such as family matters and the political changes from the 90's:

> I had to give up my job at the time because my daughter had given birth and there were no kita (nursery or daycare centre) places, so I had to help her. Then I realised how good I felt without the external pressure even though I had a nice boss at the time. I also asked my mother and found out in our family there have been many entrepreneurial persons, so maybe I have it in my genes, this freelancer spirit. My daughter is also a freelance artist and she does not want a stable job …

Alexandra may have projected the "freelance" role model for her own daughter, and thus, the generational pattern of entrepreneurship gets to be carried on. However, one could also interpret that "precarious work patterns are somehow inherited, since if the parents work in precarious circumstances, the likelihood is high that also children will do so. Alexandra justifies and rationalizes her choice by presenting it as being somehow predestined.

The "Surprised" Generation

Even more connected with the idea of aging, is the theme of "accidently" or involuntarily joining the precariat. As **Corinna** mentioned, sometimes, persons of her generation became aware of the disadvantages associated with precarious jobs only when experiencing the first signs of aging. If during youth, a precarious job can be regarded as an exciting, adventurous, a cool freelance initiative, with the passing of the years the insecurity and vulnerability of the work conditions gain in importance. For example, **Corinna** noticed that she started to experience precarity when she saw that her health got worse and that she was aging. Thinking about the future made her aware of the precarity of her work and living conditions. She talked about the "magical age of 50" as a kind of threshold when one becomes aware of aging:

> among my friends aging has become an issue, there is always one or the other who is ill and most of us are approaching 50, and this is such a 'magic number' where you start to think you are not young anymore (...) I think this idea comes from my parents' generation because my parents were already old at 50.

Youth represents a time when one can still enjoy the positive aspects of a precarious life, since there is still the idea of a future where things might be different and might eventually get better. However, once the time perspective is shrinking and health problems appear, the perception of precarity changes. This becomes the moment when precarious individuals realize, they are a "*generation who feels lost.*" The sense of loss comes from perceiving a lack of pension security and an overwhelming sense of insecurity and vulnerability in their lives.

When members of this generation started working in the 80', old age and pension security were not on their list of concerns. As **Corinna** and other participants pointed out, when she was younger, she enjoyed the advantages of self-employment such as schedule flexibility and freedom to choose when and what to work. Freedom and rebellion against a rigid work system were more important than security. Nevertheless, **Corinna's** desire for flexibility and freedom has decreased with age, while her desire for security concerning old age became prevalent. Moreover, the socioeconomic situation has changed since the 80s and 90s and re-entering secure employment with a CV of precarious jobs has become more difficult if not impossible. Thus, **Corinna** considered her own generation as a "surprised one." In her opinion, compared to the young precariat who have never known another reality, the middle-aged precariat was caught "off guard" by the socioeconomic changes in society:

> This new internship generation has it even harder than we do. For them there are no secure jobs from the start. When I started working everybody had a secure job and you had this idea that you will be doing the job for which you were prepared and will be working in the same job until you retire. For instance, you study medicine and work in the same hospital all your life. I am not an example for this, but many of my generation are. They have not left their home towns and did not change jobs. But I chose something different (...) of course, if I would have stayed with my first work contract, I could retire with 65 and have a pension of 1500 or 1800 Euro, but like this I am looking at maybe a pension of 500 Euro per month. (...) we are such a **generation, taken by surprise** by these changes. For the next generation, they know from the start they cannot rely on a state pension, so they look for alternatives while still young and my parents' generation could live very well just from the state pension.

What **Corinna** describes lends support to existing literature about the negative consequences of precarity on mental and physical health (Broding et al. 2010; Kachi et al. 2014; Ferrie 2002; Ferrie et al. 2005; Kim and von dem Knesebeck 2015). Aspects like mobility and the ability to organize one's schedule can be interpreted as positive aspects of precarity. However, these features become less important with age or lose their attractivity as time goes by. Moreover, with older age, health problems appear. Consequently, people would like more security for the times when their energy level decreases. On the one hand, physical symptoms of aging remind them of their vulnerability which stands for negative aspects of precarity. On the other hand, the vulnerability inherent to precarity makes them think about old age and its vulnerabilities. With turning 50, precarious workers reach a stage of double vulnerability where aging adds to the precarity of life in general. Thus, the vulnerabilities caused by aging (e.g., decline in bodily strength, increased vulnerability to illness, etc.) and precarity (e.g., increased stress, decreased independence, etc.) reinforce each other.

Several participants talked about how they became precarious without noticing. Getting a precarious job was not so much their own choice, but the circumstances made them end up working in precarious circumstances. For instance, **Kai** is a freelance researcher in his early fifties who works on part-time contracts and teaches at different universities in Berlin. He first came to Berlin from West Germany to study and had many dreams of an academic career. Currently he only manages to get work contracts that are with a time limit. While taking on a lot of teaching, he also needs to keep up the academic standards (e.g., publish the required number of articles or books, participate in conferences, teach) in order to be ready in case a new job opening appears. **Kai** is married and has two children. His wife also works as a freelancer, so they struggle to make ends meet. Saving money for pension does not seem very realistic at the moment. When asked about how he represented his old age, **Kai** talked about all his current worries. These included having a small pension, not being able to afford to pay a rent in old age, not getting work in the future as an old researcher or lecturer. Similar to **Corinna**, he depicted precarity as a "brutal problem" for their generation that was not prepared for working and living in such circumstances:

> The question is if I will have a stable job before I retire, and then if I really afford to retire with 65? even with the most optimistic calculations, now the pension I would get would not be enough to live on. And then the question is who will hire me as an old lecturer? Will they need me? My pension will be around 900 euro, and this will not be enough to pay the rent maybe with the inflation and all that will happen in the next twenty years. So, I will not be able to pay the rent. It will be a brutal problem for my generation. Those of us who have some financial possibilities, they take credits and buy apartments, so that they know they will not end up on the street in their old age, they will not have to sleep under a bridge one day.

Guy Standing in his book "*The precariat: the new dangerous class*" (2011) mentions the conflict between generations. Standing (2011) suggests that young precarious workers will not be able to care for their aging parents. Lending support to this idea, **Corinna's** situation shows that the middle-aged precarious generation is not really prepared to care for their parents. Nevertheless, **Corinna** mentioned that her parents still belong to the generation who have "real" pension plans and thus no real financial worries for old age.

Conflict situations may occur between the future generations of precarious individuals. Problems may arise with the younger generation since the precarious middle aged could perceive the young as competitors on the labor market. The young precariat has youth and health on their side, while middle-aged precarious persons have more experience and sometimes wisdom. However, experience and wisdom may be of little use in the dynamic job market. In order to find a job, a person needs to be learn new things and adapt quickly and experience is not so valued. The young precariat as described by Guy Standing (2011) or by Ann Allison (2013) face uncertainty on the job market from when they start being adults. They are called the "internship generation." Compared to them, the precarious middle-aged individuals have made a choice to enter self-employment not knowing that this will one day lead to them becoming part of the aging precariat. At a microindividual level, they were rebelling against the models of their parents or rigid rules of the work world (e.g., fixed time schedule, working in the same place, no freedom of movement, etc.). At a macrolevel, the liberalization of markets was encouraging self-employment, part-time jobs, and freelancing. These precarious middle agers represent *the "startled" generation* who did not save for old age. Just like in La Fontane's fable, they enjoyed the benefits of their freedom without thinking of the future. Just like the cricket facing the approach of winter, they are left without savings for their late life.

Given the disadvantages of precarity, one wonders what motivates people to hang on to precarious work conditions? Apart from rebelling against the parents' generation and breaking with conventions of a rigid work system, are there any other perceived advantages concerning precarious work? Moreover, if joining the precariat is not that difficult, does re-entering full-time, secure employment constitute a challenge? Such questions led to a closer examination of the strategies that people use in order to survive precarity.

Who Survives Precarity?

After looking at how middle-aged Germans have entered the precariat and how they perceive precarity with its positive and negative aspects, I asked myself how these people cope with precarity? I was particularly interested in those psychological traits that help people deal with the precarity of their everyday lives or the precarity of the aging experience. From the talks with middle-aged precarious Germans, several problems associated with precarity emerged. For example, the anxiety inherent in daily living was depicted as experiencing "a mental void." This "mental void" signifies that a person cannot see a future (i.e., the feeling of "no future") and has worries concerning getting by on a daily basis. The "mental void" represents the expression of perceived insecurity of living conditions as well as personal vulnerability in dealing with uncertainty.

Another issue that was often mentioned was the chaotic pace of their work days, the lack of organization or structure that characterize precarious work. A stable and secure work contract provides a structure and allows one to have hobbies and organize life around work. On the contrary, precarious work takes up a lot of time (e.g., while working on a project, a person has to look for other new projects to secure the future job). Precarious work takes up the organization of everyday life. For some people, lacking the clear structures of a stable job leaves them without orientation and sometimes with no sense of meaning in life.

In what concerns dealing with precarity, solutions included "making courageous life choices" and "constantly reinventing oneself" as well as engagement for social causes that can provide meaning in precarious circumstances.

The Chameleon Self

As depicted in the previous section, for **Alexandra**, making courageous choices in life is crucial. The way she deals with her insecure work circumstances is by telling herself one needs to have the courage to withstand the "mental void." By mental void she means periods of time when one does not know what will happen or cannot predict the future turn of events. As discussed above, "mental void" is the expression of precarity. Experiencing the "mental void" means to be aware of the inherent insecurity of the situation or one's own vulnerability. When one is courageous and can face the "mental void," then one can survive precarity. To make her point, **Alexandra**, gave the example of a famous East German actor who was offered a large amount of money for a role in a Tatort (i.e., popular German detective series). The actor badly needed money, but he did not want the part in the detective series. Even though he would have needed the money, he had the courage to withstand the "mental void" associated with being out of work for a while. In the end, it turned out all right because he got a much better offer that he would have refused, had he worked for the Tatort.

Alexandra acknowledges and accepts the insecurity inherent to her work life which she describes as the "mental void." She is aware she needs to learn to deal with this "mental void" in order to survive in precarious times. Her own coping mechanism for facing the "mental void" is constantly reinventing herself. This is what she calls her "chameleon self." In her view, it gives her more chances on the job market and helps her deal with the feeling of insecurity. Dealing with the "mental void" also requires self-efficacy, a feeling of self-confidence that one would succeed in attaining ones' goals no matter what challenges come (Bandura 1997). When people have high general self-efficacy, they believe they can manage whatever situations they may encounter. They have faith in themselves and their capacities to cope with whatever life offers. Thus, self-efficacy, (i.e., believing that one can solve any problem that one may encounter) becomes a very relevant resource in a precarious work context. The "mental void" can manifest itself in feelings of despair, but Alexandra believes it is important to learn to deal with these:

> The important thing is not to despair, when work does not go well, as it happens for many freelancers, you should not despair and give up

Alexandra talked about how she is aware she cannot stop working is she wants to afford a decent living. The competition is fierce within her work field and also among precarious workers. She sees her situation as being different from the middle-aged Germans who are originally from West Germany. However, there are similarities such as lack of secure pension and dealing with a competitive work market. **Alexandra** is motivated by the standard of living she wants to have for herself:

> Since I lived part of my life in the East, I had worked there and payed the pension fund. But I did not make any savings and did not think about my pension years. I did have some savings, but when my daughter was ill, all this money was then gone. And the insurances here in the West like the life insurance came all a bit late for me. But this is also my motivation to keep working. I cannot afford to rot and not work, because I want to have enough money to buy a coffee when I want to and not to have to count every penny I have.

For **Alexandra**, the time to stop working will come when she does not get hired anymore and not with the start of a retirement since she cannot afford to retire. Also, **Alexandra** is aware of the other threats of precarity such as not having money in old age (i.e., financial precarity) or not being healthy (i.e., frailty of the body). When one is not hired or not fit for a job anymore, it does not mean one stops working. In **Alexandra's'** view that is the time when one starts to think what one can do differently. It is time to reinvent oneself once more. Being dependent on others in old age or being ridiculous at one's job would be embarrassing:

> If I am ill, then things change, but as long as I am healthy and fit I will work. I find it also awful how some people in my generation are so fixed on the pension. It is a hard job being an interpreter, and when you get older, your voice changes and you cannot work as much, your voice is a bit strident. Then it is time to step down. My daughter will tell me when my voice is not good enough and I am also very self-critical. It is anyway a competitive job, so my colleagues will tell me anyway, or the clients just do not hire me anymore. Then you have to stop and rethink your work, do something else, it should not get embarrassing.

Alexandra believes it is very important that in precarious circumstances people develop, change, and adapt themselves in order to survive. That is why, I have called this "*the chameleon self*" the ability to redefine oneself to adapt to new circumstances and by doing so survive precarity:

> I define myself through myself. I believe the artists call it reinvention of the self ...I do this every year, for Christmas I take some time for myself and I reflect what I have learned, what I want to change about myself. Because of the nature of my work I am obliged to constantly develop ...I also seek advice from my family or from professionals

The Social Activist

Past studies show that unemployed individuals have trouble finding a structure to their days and meaning outside work. Unemployed people are known to have worse mental health and physical health then fully employed individuals (Toge and Blekesaune 2015). Among the explanations found for this phenomenon are the lack of structure and meaning that unemployed people experience in their daily lives. Health problems among unemployed people are usually the result of engaging in risk behavior like alcohol consumption or gambling (Reine et al. 2013).

In similar ways, precarity can convey the feeling of lack of meaning since one is constantly on the run and has no time to enjoy work or life in general. Days go by with the feeling of not having achieved much while being under the constant pressure to achieve more. Moreover, precarity can generate feeling of loneliness associated with the fact that one does not belong to a group. Since the precariat is not a clearly defined class, it is difficult to come together as a group. For example, when working only for a limited time or part time for an institution, people have little time to socialize with colleagues or may find it irrelevant since they "have to move" soon. Even when they recognize themselves as precarious, individuals may not want to spend time with other members of the precariat who are in competition for the same resources. Thus, a critical issue regarding precarious individuals is how to help them find meaning in their precarious work life and help them foster social relations. One potential solution is engaging in volunteer work, as one of the study participants recounted during the interviews.

Thomas is in his early fifties and is currently trying to make ends meet as a freelancer. Before trying out as a freelancer, he has been unemployed for several years. When he first started working, he used to have a well paid and secure job. However, he decided to give up his job because it did not agree with his core beliefs. After he quit his job, **Thomas** came to Berlin in the early nineties and enrolled at one of the main universities. He studied political science but then did not find any job in this domain and was unemployed for several years. Nevertheless, Thomas is very active in several non-governmental organizations with ecological and social missions, especially one for saving the rainforest. Thus, through his volunteer activities, his days are filled with sense and significance. He goes to meetings, to demonstrations, and gathers signatures for different projects. He proudly told me how gathering many signatures is a job in itself. Through his volunteer activities, he also meets people with similar values and thus his social network expands.

Thomas's story shows us that precariousness must not be automatically associated with anomie and alienation. Volunteering can also provide meaning in life and can also represent work where one applies their knowledge. Of course, this is an example of a middle-aged person and this kind of volunteering might become more difficult in old age (i.e., the problem that youthful activities may not be available or appropriate for the old-old). Nevertheless, studies show that volunteer activities can maintain if not improve health over the years. For example, one study on elderly found that volunteer activities can decrease the mortality risk. However, the motives for volunteering are relevant. Volunteering for others (others-oriented as opposed to self-oriented reasons) can contribute to a rise in longevity (Konrath et al. 2012). This is surely **Thomas's** case because he likes to help others. For example, he had an older woman friend whom he was visiting regularly and helping:

> She was my friend, she was 85 years old and she had a heart attack so she had to use a wheelchair to move. I liked to help her and in the last few weeks we were on tour around the city, so to speak…. she is now dead, and I am very upset. I am also upset because of her situation in her last 15 years. She was really one of the women who helped build Berlin again after the war, but in her pension years she lived in poverty.

Previous research has shown that volunteers have lower rates of depression (Lum and Lightfoot 2005), better daily functioning, and psychological well-being (Greenfield and Marks 2004; Piliavin and Siegl 2007; Thoits and Hewitt 2001). Volunteering contributes to a sense of deeper meaning, called eudaimonic well-being, as compared to hedonistic well-being (Piliavin and Siegl 2007). Other researchers suggested that volunteering can prevent feelings of anomie (Musick et al. 1999). Thus, for precarious individuals who engage in volunteering, this can act as a buffer against feelings of anomie.

Apart from its psychological and health benefits, volunteering provides the opportunity to protest against social injustice and attempt to make some changes. **Thomas** criticized the system and the current economic situation. He mentioned the high numbers of unemployed academics in Berlin who have maybe PhD or several PhDs and still have no work or have to take menial jobs such as his friend with two PhDs who works as a cleaner. **Thomas** criticized the condition of the "*precarious intellectual,*" forced to take up jobs that are under their education level. The drama of such persons is that they are overqualified for many potential jobs, and their education level does not constitute an asset for finding work. **Thomas** is very much concerned with doing something for society and for finding solutions to social and economic problems. The reasons he gave up his comfortable job in the first place is because he did not agree with the implications of his work. He also wanted to do more for society in general. Similar to other precarious workers, **Thomas** has realized that the neoliberal ideology wants to encourage individualism and thus make people more competitive and lonelier. However, through social engagement, people come together and fight for the same causes. This brings again a feeling of belonging and helps to cope with precarity since the latter represents a feeling of being vulnerable and lonely.

Thomas does not want to stay unemployed. For his future, **Thomas** tries to reinvent himself professionally and become self-employed. He does not have hopes for a stable job that would fit his interests and motivations. However, he can make use of his skills and teach as a freelance lecturer.

> I try to find work and get out of unemployment, I want to become self-employed but of course it is a difficult path…It is a reflection of our times where people are encouraged to invent their own jobs, this is advertised in the media and presented as a wonderful chance to make your career dreams come true. The reality of course is harsher because you are being pushed into self-employment and this means you are responsible for all your insurances and pension

In his desire to become a freelancer, **Thomas** may have been indirectly influenced by the general media discourse that encourages people to enter self-employment and build their own "success story."

All in all, one must take into consideration the fact that **Thomas** has time to volunteer because he is unemployed at the moment. Things might change for him when he becomes a freelancer and thus "regains" the full status of "precarious worker." Nevertheless, at present **Thomas** has created a mission for himself to improve the lives and work conditions of precarious workers and he is a firm believer in the introduction of a national statutory minimal wage and minimal pension.

References

Allison, A. (2013). *Precarious Japan*. Duke University Press. http://dx.doi.org/10.1215/9780822377245.

Arber, S. & Ginn, J. (2004). Ageing and gender: Diversity and change. *Social Trends, 34,* 1–14. London: The Stationery Office.

Axelsson, L., Malmberg, B., & Zhang, Q. (2017). On waiting, work-time and imagined futures: Theorising temporal precariousness among Chinese chefs in Sweden's restaurant industry. *Geoforum, 78,* 169–178. https://doi.org/10.1016/j.geoforum.2015.12.007.

Bandura, A. (1997). *Exercise of personal and collective efficacy in changing societies. Self-Efficacy in Changing Societies*, 1–45. https://doi.org/10.1017/cbo9780511527692.003.

Bäcker, G. (2007) 'Was heißt hier "geringfügig"? Minijobs als wachsendes Segment prekärer Beschäftigung' [What does it mean „small"? Minijobs and the rising number of precarious jobs]. In B. Keller and H. Seifert (eds) *Atypische Beschäftigung – Flexibilisierung und soziale Risiken [Atypical work. Flexibilisation of work and social risks]*. Berlin: Sigma.

Buntenbach, A. (2007). *Prekäre Beschäftigung, Herausforderung für die Gewerkschaften. Anregungen und Voschläge für die gewerkschaftliche Diskussion [Precairous jobs. Challenges for Unions and Recommendations for Union Discussions]*. Düsseldorf: Deutscher Gewerkschaftsbund.

Bardasi, E., & Jenkins, S. P. (2010). The gender gap in private pensions. *Bulletin of Economic Research, 62*(4), 343–363. https://doi.org/10.1111/j.1467-8586.2009.00336.x.

Bardasi, E., Jenkins, S., & Rigg, J. (2002). Retirement and the income of older people: A British perspective. *Ageing & Society, 22,* 131–159.

Bispinck, R., & Schulten, T. (2011). *Trade Union Responses to Precarious Employment in Germany. German Report to the Project "Bargaining for Social Rights"* (BARSORI) financed by the European Commission. Retrieved from https://www.econstor.eu/handle/10419/52723.

Biegert, T. (2014). On the outside looking in? Transitions out of non-employment in the United Kingdom and Germany. *Journal of European Social Policy, 24*(1), 3–18.

Brady, D., & Biegert, T. (2017). The rise of precarious employment in Germany. *Research in the Sociology of Work*, 245–271. https://doi.org/10.1108/s0277-283320170000031008.

Brady, D., Fullerton, A. S., & Cross, J. M. (2010). More than Just nickels and dimes: A cross- national analysis of working poverty in 18 affluent democracies. *Social Problems, 57*(4), 559–585.

Broding, H. C., Weber, A., Glatz, A., & Bünger, J. (2010). Working poor in Germany: Dimensions of the problem and repercussions for the health-care system. *Journal of Public Health Policy, 31*(3), 298–311. https://doi.org/10.1057/jphp.2010.20.

Buffel, T., & Phillipson, C. (2011). Experiences of place among older migrants living in inner-city neighbourhoods in belgium and england. *Diversité Urbaine, 11*(1), 13. https://doi.org/10.7202/1007742ar.

Bührmann, A. D. & H. J. Pongratz (Eds.). (2010). *Prekäres Unternehmertum. Unsicherheiten von selbstständiger Erwerbstätigkeit und Unternehmensgründung*. [Precarious Initiative. Insecurity for freelance activitites and initiatives]. Wiesbaden: VS Verlag.

Burda, M. C. (2016). The German labor market miracle, 2013–2015: An assessment. SFB 649 Discussion Paper 2016-005, Humboldt-Universität zu Berlin.

Bundesagentur für Arbeit. (2011): *Beschäftigung in Deutschland [Work in Germany]*: Stichtag 31. März 2011, Nürnberg.

Buchholz, S., & Blossfeld, H.-P. (2009). Beschäftigungsflexibilisierung in Deutschland –Wen betrifft sie und wie hat sie sich auf die Veränderung sozialer Inklusion/Exklusion in Deutschland ausgewirkt? [The flexibilisation of work in Germany-Who is affected and how did it influence the inclusion/exclusion in Germany?] In R. Stichweh & P. Windolf (Eds.), *Inklusion und Exklusion: Analysen zur Sozialstruktur und sozialen Ungleichheit [Inclusion and Exclusion: Analysis of the Social Structure and Social Inequality]*. 123–138. Wiesbaden: VS Verlag für Sozialwissenschaften.

Buchholz, S., & Grunow, D. (2006). Women's Employment in West Germany. In H.-P. Blossfeld & H. Hofmeister (Hrsg), *Globalization, Uncertainty and Women's Careers. An International Comparison* (S. 61–83). Cheltenham: Edward Elgar Publishing.

Bylow, C. & Vaillant, K. (2014). *Die verratene Generation Was wir den Frauen in der Lebensmitte zumuten.* [The betrayed generation. What to expect from middle-aged women?] Pattloch.
Crăciun, I. C. (2018). Aging in Precarious Times. *Alter Und Geschlecht*, 163–175. https://doi.org/10.1007/978-3-658-20872-1_11.
Dannefer, D. (2003). Cumulative advantage/disadvantage and the life course: Cross-Fertilizing age and social science theory. *The Journals of Gerontology Series B: Psychological Sciences and Social Sciences, 58*(6), S327–S337. https://doi.org/10.1093/geronb/58.6.s327.
Engstler, H., & Romeu Gordo, L. (2016). Der Übergang in den Ruhestand: Alter, Pfade und Ausstiegspläne. [Transition to Pension: Aging, Paths and Plans] In Mahne et al. (Eds.) *Altern Im Wandel* [Changes in Aging], 65–80. https://doi.org/10.1007/978-3-658-12502-8_4.
Euronews. (2017). Berlin is the capital of the working poor. Retrieved from https://www.euronews.com/2017/07/26/dierk-hirschel-berlin-.
EURES. (2018). The European job mobility portal. Retrieved from https://ec.europa.eu/eures/main.
Ferraro, K. F., & Shippee, T. P. (2009). Aging and cumulative inequality: How does inequality get under the skin? *The Gerontologist, 49*(3), 333–343. https://doi.org/10.1093/geront/gnp034.
Ferrie, J. E., Shipley, M. J., Newman, K., Stansfeld, S. A., & Marmot, M. (2005). Self-reported job insecurity and health in the Whitehall II study: Potential explanations of the relationship. *Social Science and Medicine, 60*(7), 1593–1602. https://doi.org/10.1016/j.socscimed.2004.08.006.
Ferrie, J. E. (2002). Effects of chronic job insecurity and change in job security on self reported health, minor psychiatric morbidity, physiological measures, and health related behaviours in British civil servants: the Whitehall II study. *Journal of Epidemiology and Community Health, 56*(6), 450–454. https://doi.org/10.1136/jech.56.6.450.
Foster, L., & Walker, A. (2014). Active and successful aging: A European policy perspective. *The Gerontologist, 55*(1), 83–90. https://doi.org/10.1093/geront/gnu028.
Gallie, D., Paugam, S., & Jacobs, S. (2003). Unemployemnt, poverty and Social Isolation: Is there a vicious circle of social exclusion? *European Societies, 5*(1), 1–32. https://doi.org/10.1080/1461669032000057668.
Geyer, J., & Steiner, V. (2010). Public pensions, changing employment patterns, and the impact of pension reforms across birth cohorts - A microsimulation analysis for Germany. *SSRN Electronic Journal*. https://doi.org/10.2139/ssrn.1577549.
Greenfield, E. A., & Marks, N. F. (2004). Formal volunteering as a protective factor for older adults' psychological well-being. *The Journals of Gerontology, Series B, 59,* 258–264. https://doi.org/10.1093/geronb/59.5.S258.
Han, K.-M., Chang, J., Won, E., Lee, M.-S., & Ham, B.-J. (2017). Precarious employment associated with depressive symptoms and suicidal ideation in adult wage workers. *Journal of Affective Disorders, 218,* 201–209. https://doi.org/10.1016/j.jad.2017.04.049.
Harkness, S. (2002). *Low pay, times of work and gender*. Manchester: Equal Opportunities Commission.
Hohendanner, C. (2010): Unsichere Zeiten, unsichere Verträge? Befristete Arbeitsverträge zwischen Auf- und Abschwung, IAB-Kurzbericht No. 14. [Uncertain times, uncertain contracts? Temporary work contracts between upswing and upturn. Short report of IAB, nr. 14]. Retrieved from http://doku.iab.de/kurzber/2010/kb1410.pdf.
Kachi, Y., Otsuka, T., & Kawada, T. (2014). Precarious employment and the risk of serious psychological distress: A population-based cohort study in Japan. *Scandinavian Journal of Work, Environment & Health, 40*(5), 465–472. https://doi.org/10.5271/sjweh.3442.
Kalleberg, A. L. (2013). Book review symposium: Response to reviews of Arne L Kalleberg, Good Jobs, Bad Jobs: The rise of polarized and precarious employment systems in the United States, 1970s to 2000s. *Work, Employment & Society, 27*(5), 896–898. https://doi.org/10.1177/0950017013480403.
Kalleberg, A. L. (2009). Precarious work, insecure workers: Employment relations in transition. *American Sociological Review, 74*(1), 1–22. https://doi.org/10.1177/000312240907400101.
Kalleberg, A. (2011). *Good Jobs, Bad Jobs: The rise of polarized and precarious employment systems in the United States, 1970s to 2000s*. New York: Russell Sage.

Kim, I. H., Muntaner, C., Vahid Shahidi, F., Vives, A., Vanroelen, C., & Benach, J. (2012). Welfare states, flexible employment, and health: A critical review. *Health Policy, 104*(2), 99–127. https://doi.org/10.1016/j.healthpol.2011.11.002.

Kim, T. J., & von dem Knesebeck, O. (2015). Is an insecure job better for health than having no job at all? A systematic review of studies investigating the health-related risks of both job insecurity and unemployment. *BMC Public Health, 15*(1). https://doi.org/10.1186/s12889-015-2313-1.

Kock, K. (Ed.). (2008). *Der Preis der Freiheit. Solo-Selbstständige zwischen Vermarktung, Professionalisierung und Solidarisierung. [The price of freedom. Freelancers between marketing, professionalising and solidarising].* Dortmund: Sozialforschungsstelle.

Konrath, S., Fuhrel-Forbis, A., Lou, A., & Brown, S. (2012). Motives for volunteering are associated with mortality risk in older adults. *Health Psychology, 31*(1), 87–96. https://doi.org/10.1037/a0025226.

Kelle, N., Simonson, J., & Romeu Gordo, L. (2012). Veränderte Erwerbsverläufe von Männern in Ost- und Westdeutschland: Ein Vergleich der Babyboomer mit älteren Kohorten [Altered career paths of men from East and West Germany: A comparison of Babyboomers with older cohorts]. *Ifo Dresden berichtet, 19*(3), 24–29.

Lum, T. Y., & Lightfoot, E. (2005). The effects of volunteering on the physical and mental health of older people. *Research on Aging, 27,* 31–55. https://doi.org/10.1177/0164027504271349.

Malenfant, R., LaRue, A., & Vézina, M. (2007). Intermittent Work and Well-Being. *Current Sociology, 55*(6), 814–835. https://doi.org/10.1177/0011392107081987.

Min, K.-B., Park, S.-G., Hwang, S. H., & Min, J.-Y. (2015). Precarious employment and the risk of suicidal ideation and suicide attempts. *Preventive Medicine, 71,* 72–76. https://doi.org/10.1016/j.ypmed.2014.12.017.

Mahne, K., Wolff, J. K., Simonson, J., & Tesch-Römer, C. (2016). Altern im Wandel: Zwei Jahrzehnte Deutscher Alterssurvey. Altern Im Wandel, 11–28. https://doi.org/10.1007/978-3-658-12502-8_1.

McDowell, L. (2013). *Working lives: Gender, migration and employment in Britain, 1945–2007.* Chichester: Wiley-Blackwell.

McNamara, M., Bohle, P., & Quinlan, M. (2011). Precarious employment, working hours, work-life conflict and health in hotel work. *Applied Ergonomics, 42*(2), 225–232. https://doi.org/10.1016/j.apergo.2010.06.013.

Musick, M., Herzog, A. R., & House, J. S. (1999). Volunteering and mortality among older adults: Findings from a national sample. *Journals of Gerontology. Series B, Psychological Sciences and Social Sciences, 54,* S173–S180. https://doi.org/10.1093/geronb/54B.3.S173.

Naumann, D., & Romeu Gordo, L. (2010). Gesellschatliche Partzipazion: Erwerbstäti gkeit, Ehrenamt und Bildung. In A. Motel-Klingebiel, S. Wurm & C. Tesch-Römer (Hrsg.) *Altern im Wandel. Befunde des Deutschen Alterssurveys (DEAS)* (S. 118–141). Stuttgart: Kohlhammer.

Piliavin, J. A., & Siegl, E. (2007). Health benefits of volunteering in the Wisconsin Longitudinal Study. *Journal of Health and Social Behavior, 48,* 450–464. https://doi.org/10.1177/002214650704800408.

Price, D. (2006). Why are older women in the UK poor? *Quality in Ageing and Older Adults, 7*(2), 23–32. https://doi.org/10.1108/14717794200600011.

Portacolone, E. (2013). The notion of precariousness among older adults living alone in the U.S. *Journal of Aging Studies, 27*(2), 166–174. https://doi.org/10.1016/j.jaging.2013.01.001.

Portacolone, E. (2017). Structural factors of elders' isolation in a high-crime neighborhood: An in-depth perspective. *Public Policy & Aging Report, 27*(4), 152–155. https://doi.org/10.1093/ppar/prx025.

Reine, I., Novo, M., & Hammarström, A. (2013). Unemployment and ill health – A gender analysis: Results from a 14-year follow-up of the Northern Swedish Cohort. *Public Health, 127*(3), 214–222. https://doi.org/10.1016/j.puhe.2012.12.005.

Reynolds, J., & Wenger, J. B. (2010). Prelude to a RIF: Older workers, part-time hours, and unemployment. *Journal of Aging & Social Policy, 22*(2), 99–116. https://doi.org/10.1080/08959421003620715.

Simonson, J., Gordo, L. R., & Kelle, N. (2015). Separate paths, same direction? Destandardization of male employment biographies in East and West Germany. *Current Sociology, 63*(3), 387–410. https://doi.org/10.1177/0011392115572380.

Simonson, J., Gordo, L. R., & Titova, N. (2011). Changing employment patterns of women in Germany: How do baby boomers differ from older cohorts? A comparison using sequence analysis. *Advances in Life Course Research, 16*(2), 65–82. https://doi.org/10.1016/j.alcr.2011.03.002.

Simonson, J., Kelle, N., Romeu Gordo, L., Grabka, M. M., Rasner, A., & Westermeier, C. (2012). Ostdeutsche Männer um 50 müssen mit geringeren Renten rechnen. *DIW Wochenbericht, 79*(23), 3–13.

Simonson, J. (2013). Erwerbsverläufe im Wandel – Konsequenzen und Risiken für die Alterssicherung der Babyboomer [Changing work tranjectories. Consequences and risks fort he insurance oft he baby boomers]. In C. Vogel & A. Motel-Klingebiel (Eds.). *Altern Im Sozialen Wandel: Die Rückkehr Der Altersarmut? [Aging and Social Change: The Return of Poverty in Old Age?]*, 273–290. https://doi.org/10.1007/978-3-531-18714-3_14.

Schaefer, A., & Gottschall, K. (2015). From wage regulation to wage gap: How wage-setting institutions and structures shape the gender wage gap across three industries in 24. European Countries and Germany. *Cambridge Journal of Economics, 39*, 467–496.

Smith, A., Sim, J., Scharf, T., & Phillipson, C. (2004). The quality of life in deprived neighbourhoods. *Ageing & Society, 24*, 814–937.

Standing, G. (2011). *The precariat: The new dangerous class*. London: Bloomsbury Academic.

Statistisches Bundesamt. (2011). Mikrozensus 2010, Bevölkerung und Erwerbstätigkeit. Beruf, Ausbildung und Arbeitsbedingungen der Erwerbstätigen in Deutschland. [Census 2010, Population and work. Jobs, internships and work conditions for workers in Germany]. *Fachserie 1, Reihe 4.1.2*. Wiesbaden: Statistisches Bundesamt.

Unger, B. (2015). *The German model: Seen by its neighbors*. Düsseldorf, Germany: Institute of Economic and Social Research.

Thelen, K. (2012). Varieties of capitalism: Trajectories of liberalization and the new politics of social solidarity. *Annual Review of Political Science, 15*, 137–159.

Thompson, S. (2013). *States of uncertainty: Youth unemployment in Europe*. London: Institute for Public Policy Research.

Toge A. G. & Blekesaune M. (2015). Unemployment transitions and self-rated health in Europe: A longitudinal analysis of EU-SILC from 2008 to 2011, *Social Science & Medicine, 143*, 171–178.

Thoits, P. A., & Hewitt, L. N. (2001). Volunteer work and well-being. *Journal of Health and Social Behavior, 42*, 115–131. https://doi.org/10.2307/3090173.

Vitols, S. (2004). Continuity and change: Making sense of the German model. *Competition & Change, 8*(4), 331–337.

Weinkopf, C. (2009). Germany: Precarious employment and the rise of mini-jobs. Retrieved from https://www.researchgate.net/publication/261358207.

Ziegelmeyer, M. (2013). Sind Selbstständige von Altersarmut bedroht? Eine Analyse des Altersvorsorge-Verhalten von Selbstständigen. [Are freelancers at risk for poverty in old age? An analysis of the insurance behaviour of freelancers] In C. Vogel & A. Motel-Klingebiel (Eds.), *Altern Im Sozialen Wandel: Die Rückkehr Der Altersarmut?* [Aging and Social Change: The Return of Poverty in Old Age?] 229–249. https://doi.org/10.1007/978-3-531-18714-3_12.

Chapter 6
Views on Aging in Precarious and Financially Secure Individuals

Introduction

Views on aging emerge in a social context and are shared by individuals with similar life circumstances (Fung 2013; Wurm and Huxhold 2012; Wolff et al. 2018). Therefore, in an attempt to better understand aging representations emerging within a precarious context, I set out to compare precarious middle-aged individuals and their counterparts, people with secure jobs and pension plans. The sample of ***financially secure middle-aged Germans*** included women and men aged between 45 and 65. In contrast to the aging precariat, these participants had stable jobs, a secure pension plan, or owned financial assets such as property. They all lived in Berlin at the time when the research was conducted, but some were originally from other West German cities or were immigrants.

As mentioned in the introduction, the interview method used (i.e., episodic interviews) helps to elicit information on both representations of aging or lay theories that individuals develop about aging (i.e., semantic information) and the contexts where they develop these (i.e., episodic knowledge). The latter are relevant because they represent situations that would trigger either positive or negative thoughts on aging and may eventually help in the process of changing these if necessary. Views on aging were investigated through questions about how participants define old age and aging and when they consider a person to be old. During the interviews, participants were first asked if they think about aging, and if yes in which situations they think about old age or aging (i.e., triggers of automatic thoughts about aging). Next, interviewees were invited to describe an old person and give their opinion on when one can start calling a person "old" or "senior." Last but not least, participants had to define old age from their own perspective.

The chapter aimed to depict the views on aging, that are inherent to a precarious reality, characterized by demographic changes and financial uncertainty, that both precarious and financially secure individuals share. The circumstances for thinking

I. C. Crăciun, *Positive Aging and Precarity*, International Perspectives on Aging 21, https://doi.org/10.1007/978-3-030-14255-1_6

about aging are described and compared in the first half of the chapter. In the second half of the chapter, definitions, and views on aging of precarious and financially secure individuals are presented and contrasted.

Contexts that Trigger Thoughts About Aging

In order to better understand how views on aging emerge in a social context, participants were asked to describe typical situations when they think about aging. Answers comprised a variety of personal and social circumstances. The prompts for thinking about aging were very similar for precarious and financially secure participants. For example, comparison with either the older generation (e.g., parents, grandparents, older friends) or the younger generation (e.g., children, young colleagues, or a younger version of oneself) constituted a frequent stimulus for thoughts concerning aging. Nevertheless, comparisons were usually negative in the sense that youth was associated with positive traits, while the aging self with negative characteristics. On the one hand, participants compared their present self with their former youthful self and noticed the effects of aging. These aging induced changes were described in terms of decline. On the other hand, middle-aged individuals compared themselves with the younger generation and perceived how the latter performed better on work tasks or "just looked better." Likewise, aging thoughts emerged when participants noticed aging signs in their friends or their parents. These thoughts were expressed mostly in terms of worries about the health of parents or where they will live when their health will not permit them to be independent anymore (e.g., what are the options for an institution or hiring someone to take care of their old parents).

Fifty Is a Magic Number

One of the most often mentioned situations when people find themselves thinking about aging was when they experienced "aging symptoms." Such symptoms were pains or aches that or changes in appearance (e.g., gray hair, unwanted extra pounds, etc.) reminded them that they are "getting older." Because they were associated with a negative view on old age and old people, such thoughts usually generated anxiety about the future.

If they had to name a certain age when aging thoughts start to be more present, participants said that this would be the moment when they turn or will be fifty years old. For instance, **Corinna**, a freelancer business woman, mentioned that ***turning fifty*** represented a defining moment when suddenly aging becomes a concern that is shared with friends. Aging represents an issue that participants often discussed with their peers who were the same age because they knew that their friends would understand. For instance, in **Corinna's'** friends' circle, aging:

become a topic for us all. One or the other is ill, and we are all getting older. We are all around 50 years old, and **50 is like a magic number**, you know you are not that young anymore…

Thus, turning fifty is regarded as the "magic threshold" for contemplating old age. Even if precarious individuals ignored aging thoughts, after they turned fifty, the awareness of aging signs made old age more difficult to ignore.

Similar to Corinna, **Daniel**, a freelance musician in his late fifties, considered that old age is something one begins to contemplate later in life. In his view, one starts to think about aging when the physical signs cannot be denied anymore:

> Aging is for me a topic that I think is usually denied, not thought about, I also haven't thought so much about it, but now I realise I should

As mentioned above, aging was also noticed when comparing oneself with others. For example, seeing their parents grow old or noticing aging symptoms in their friends made most participants apprehensive about their own aging process. For instance, **Kai**, a freelance researcher in his early fifties talked about how aging signs in his parents and siblings reminded him of his own aging and made him pay more attention to the physical signs he experiences himself:

> I think about aging when I see my parents' generation, I notice their health decline, I also have siblings that are around 10–15 years older than me and I can observe what awaits me. I also see for myself that I am getting slower and things are not as they used to be, maybe I need some intensive treatment at the dentist or a new pair of glasses, and all these things cost money, it is very different compared to the 80s…(…) I also notice the passage of time I am aware how the time is passing faster than when I was younger, I used to believe I have a lot of time to deal with things, so this is now a great change. And of course, I fear health decline. I hope not to be dependent on care, to be a nursing home case in old age!

All in all, the prospect of poverty in old age constituted a recurrent concern for precarious middle-aged participants. Aging and poverty were often mentioned as an issue they discussed with their other precarious friends. Thinking about aging triggered automatic thoughts of poverty threat and the risk of becoming a "social case" in old age.

On the whole, reflections on aging frequently occurred in contexts where individuals encountered problems (e.g., parents' illnesses, personal health complaints, noticing aging signs in their friends, difficulties at work because of older age, anxiety about the future). Such contexts are indicative of a negative view on aging where getting older is associated with illness, health decline, loss of productivity, or poverty. The situations where people think about aging may play a relevant role for the type of thoughts that people have concerning old age. Negative contexts such as poverty, illness, or disability are more likely to activate negative thoughts about aging. In this sense, contexts that reflect precarity, either of the body (e.g., health decline, frailty), economic (e.g., poverty, being dependent on others), or social (e.g., loneliness) can trigger negative views on aging. This activation of negative aging stereotypes happens without the persons' awareness, as described in the stereotype embodiment theory (Levy 2009) and could make negative views on aging resistant to change.

Self-definitions and Perceptions of Aging

When does old age start and how can we define old age? The answers to such questions are relevant for defining a person's identity. Both precarious individuals and their financially secure counterparts named physical characteristics and symptoms as relevant for "recognizing" old age. Old age was defined in terms of decline, dependency, and illness showing a negative view on aging. In the following, I will discuss the various aspects of this negative views on aging.

Negative Views on Aging

The Treacherous Aging Bodies

When asked how they identify a person as being old, both precarious and financially secure participants referred to bodily signs. The body was labeled as an "identity card" that helps one to recognize the signs of aging "*or how old a person really is.*" However, the aging body was described as being deceitful because it can both conceal and relevel the person's age. For example, **Alexandra**, a freelance interpreter in her early sixties, mentioned looking at a person's skin and hands since

> they usually give away their real age. Or sometimes you notice aging signs around the neck. With some people you cannot really tell their age! If they are still thin like my mother used to be, and she also had great skin, so it was really difficult to guess her real age

For precarious workers, being young and attractive increases the desirability on the job market. Thus, a person's physical appearance becomes the identification tool that can reveal or hide a person's true age. The question arises why is it important to hide one's true age from others? For instance, in **Alexandra's'** line of work, apart from physical appearance, the voice represents an important giveaway for a person's actual age. In her view, interpreters must keep their voice sounding young and even attractive if they want to keep their job when they get older. Showing signs of physical decline is directly connected to work performance and may mean losing ones' job:

> I have many colleagues who are still active, but their voice just sounds old. It is time for them to stop working! In my case, my daughter will tell me when I am not competent for the job anymore. I believe one needs to be very critical with oneself! Anyway, our job is tough, and the colleagues tell you when your voice is not good enough or the clients stop hiring you for interpreting jobs, it is as simple as that!

Continuing work in late life is of vital importance to self-employed middle-aged Germans like **Alexandra**. Thus, precarious individuals need to project an image of youth and productivity in order to continue to get hired. According to **Daniel**, a freelance musician in his late fifties, there is a social scale that is used to evaluate people, namely

> those who are still young and healthy in their old age and can continue working are high up in that hierarchy, and the ones who are social cases and require care, score low on that social scale. Between the top and the bottom of the hierarchy there is a huge gap…

Precarious individuals need to live with the awareness of this scale and they must fight to stay at the top of the scale in their old age. If they fail to do so, they might lose their financial means of existence and become the much dreaded "social cases" that they fear to be in their old age. However, if one manages to preserve the youthful, wholesome looks, one has a chance of being perceived as a successful ager.

Physical appearance represents an aging indicator for ***financially secure individuals*** as well. For example, gray or white hair and wrinkles were considered typical features of old age. Aging signs are often noticed in the appearance of friends, as **Charlotte**, a journalist in her mid-fifties explained:

> I am now 55 and some colleagues of mine I know already for 15 years, they are more or less the same age as me and I notice their skin, the wrinkles, how tired their faces look! Especially if I haven't seen some friends in a while, because they were away on holiday or had a six-month sabbatical, when they come back, I take in the changes. They have aged!

For the interviewed women, physical appearance is directly connected to attractiveness in old age. **Clara**, a school teacher approaching retirement, said that

> you notice you have aged because people do not find you attractive anymore, you don't find yourself as attractive as before

The ***fear of not being attractive anymore*** in old age was mentioned predominantly by interviewed women. Furthermore, it was often presented as an indicator of their own aging process, as **Britta**, a bank clerk in her early fifties, observed:

> when I look at myself in the mirror, I see I have aged…it is normal, but I also find it awful

Even if women judged their age by their image in the mirror, fear of being unappealing it in old age is not just a female preoccupation. Some financially secure participants like **Fabian**, a wealthy man, explained why he finds aging dreadful:

> Aging is not pleasant…you start noticing your eyes are not good enough, you need glasses, you start to put on weight and you panic, you start jogging but it's no use… I try to fight old age, to defend myself against it, I don't want to be old! …I can work a bit to change my appearance so that I don't look like crap! Naturally, I go to the beauty salon and I die my hair

For financially secure individuals, the aging bodyis treacherous in the sense that it betrays their desire to be youthful and attractive. It also is unsatisfactory because it is not as functional as it used to be (e.g., you cannot run as before, you cannot see as before, etc.). Thus, the aging body betrays their expectations. The body becomes "unreliable" as it ages.

Overall, both precarious and financially secure participants though it was important to hide the physical aging signs (i.e., the aging body) to prevent others from noticing ones' "*real*" age (i.e., chronological age). They did not want to be associated or identify themselves with the negative stereotype of "old person." The negative characteristics associated with old age became a barrier toward self-acceptance of the aging process.

The act of **concealment** takes individuals far from the ideal of self-acceptance and increases the gap between body (e.g., physical signs of aging) and mind (e.g., life experience). Efforts to hide ones' real age give rise and maintain the anxiety that one may be discovered, that people may guess their real age. Moreover, the energy invested in concealing ones' age can deplete resources that could be used for other activities. In case of the precarious middle-aged individuals, the fear of their real age being discovered is associated with the anxiety of losing their jobs or future hiring opportunities. For financially secure people, the dread of showing ones' real age is connected to the fear of being put into the "old age box" that is associated with ugliness and decay. Despite their worries about showing their true age, few participants mentioned actively doing something to fight aging. Except for the usual concealment of aging signs (e.g., dying ones' hair to prevent it from being gray or white) or preventive measures such as exercising to keep the body fit, no other transformations were undertaken. Nevertheless, most participants are still middle-aged and thus may not think so much about changing their appearance (e.g., cosmetic surgery) and concentrate on prevention of further decline.

Old Age Represented as an Illness

When asked to define old age, ***precarious participants*** mentioned how they first recognized aging signs while monitoring their own body. Old age is talked about as if it were an illness with its own symptoms and undesirable consequences for the body (e.g., you feel aches and pains, you are walking slower) and mind (e.g., you feel depressed, you lack motivation and enthusiasm characteristic of youth, you become forgetful and "silly").

For example, **Carmen** is a woman from a Latin American country in her early fifties. She came to Germany around 15 years ago and is currently working part-time as a freelance architect with no pension security. During the interview, she described how she noticed her first aging symptoms. For her, old age was linked to physical decline and decreased performance. Carmen mentioned both physical and psychological symptoms of old age that were associated with daily activities she could not perform as she used to. Carmen felt she was "getting slower" at carrying out her work and lacked energy for leisure pursuits:

> I believe you do not have the same strength for everyday life. When you must clean the house, cook, or do the shopping and carry heavy bags, that is when you notice you are not 20 years old anymore. This is just an example of everyday activities, but it can happen at work as well! sometimes I forget things that I would have easily remember 20 years ago. Nonetheless experience also plays a role! This is the age when my children still need me a lot, so I notice at the end of the day I am very tired. For instance, a couple of years ago I could still watch a film after midnight, and nowadays I notice at 9 pm I need to go to bed, I need the sleep because I wake up at 6 am and then I have no pause during my day. I always

> must be active during the day and in the evening, I notice I am so tired! It did not use to be like this! I used to work all day, and, in the evening, I could still go out with friends!

For ***financially secure individuals***, indicators of old age also resembled illness. Old age is pathologized, as can be recognized in the description of several physical and psychological symptoms that imply a negative representation of old age. For example, **Charlotte**, the journalist with a secure job, talked about her father and the decline he experiences while aging:

> My father is now 90 years old and during the last five years I noticed how he started to forget things. His walk is hesitant, and he often falls. He started to worry, yes, he worries.

Apart from worries, Charlottes' father also experiences changes in his personality and mental capacities. In his old age, he resembles a vulnerable child:

> With aging you become a bit more conservative, or even a bit crazy…my father gathers all these advertisements from newspapers…he believes what they promise that if you give one euro, you will get three smartphones! I tell him this is a scam, and he says I am wrong, he believes this people are so stupid and they give everything away! My father used to be a professor, so he is intelligent and knowledgeable (…) so it is this fear that you lack protection. If there is no one to take care of you, such as your children for instance, **you are easy pray for scams because you become gullible**…

The fear of dependency is linked with the fear of becoming vulnerable because of physical decline (i.e., lack of energy, illness) or mental decline in old age (i.e., one becomes gullible and forgetful), as **Charlotte** mentioned: "*it is such a big fear that you are going to be without protection…*".

Dependency was considered bad and represented the opposite of the positive image of the autonomous older person. **Fabian** expressed his concern with being dependent on an ineffective pension and health systems. Even if he himself enjoys a financially stable situation, he perceived the state system as exploitative and untrustworthy:

> to be dependent on the State…or the health system…not for me…or old people's home they are building ghettos for old people!

Fabian is financially well off and can make plans to be independent from institutional help in old age. Financial security in old age is associated with being able to decide for yourself, with independence and autonomy as **Clara** mentioned:

> in my case, I have no financial needs…my **ability to decide for myself** must stay…as long as I can decide for myself, that is security

The fear of being alone in old age was connected with the feeling of vulnerability in facing the challenges of old age and possible crises. Participants associated loneliness in old age with the feeling of not being needed by others such as family or society in general.

Precarious individuals expressed the ***fear of being unproductive*** and thus not useful in old age. For them, not being needed means not getting work and no financial security in old age. When seeking job opportunities, the older precariat enters the competition with young people and their chances of winning are small since as Alexandra said, "*younger people simply walk faster.*"

Such negative **images of decline and vulnerability**, of becoming a new, "lesser version" of oneself were described as indicators of one's personal aging process. For instance, **Britta**, a bank clerk in her early fifties, described how she became aware of her own aging process:

> I notice it becomes difficult to move, I feel vulnerable. Before I used to practice a lot of sport, and then this pain started in my hip and in my knees. I cannot go skiing as before. I notice the pain gets more severe. I have this fear not to make it worse, so I do not move my hips and knees as much as I used to! With aging my mobility is getting poorer and **this is horrible**!

Nevertheless, physical "symptoms" of aging may be perceived as less unpleasant compared to the psychological changes that happen with age. Psychological sigs of aging (e.g., decline of mental capacities, emotional changes) were mentioned especially by financially secure individuals. For instance, **Fabian** described the psychological features of getting older as being more unpleasant than the bodily changes that come with aging:

> Maybe the physical decline is not that bad, but it is bad when you notice you are older, and nothing is new and fun anymore! You become depressive, you become cynical! I became disillusioned… to be young means you can still learn things, you have not reached your final potential, but with age you lose your capacity to be motivated, to try out new things, because you have no curiosity….

On the whole, for both precarious and financially secure individuals, the beginning of old age is marked by symptoms such as pain, aches, tiredness, and the awareness of getting slower. Moreover, health problems are described as typical for old age. Identifying old age with illness represents a feature of negative images of aging (Wurm et al. 2010, 2013). Thus, we can say that both precarious and financially secure individuals have a negative idea about aging, since they associate it automatically with illness. Furthermore, just as illnesses can and should be prevented, old age also becomes preventable.

Charlotte described her female role model for positive aging. The old woman is in her eighties, but looks young because of here discipline and overall attitude toward life:

> I know this woman who lives in Paris, she does every morning 20 min of gymnastics and looks great. She is 85 but she looks 45. She rides a bike, goes swimming twice a week…. I really admire how she ages, without complaints, without grumbles"

The old woman used to work as a teacher but is retired now and enjoys a secure pension. Her schedule allows her to perform her gymnastics routine every day to maintain her fitness level and youthful looks. What **Charlotte** illustrates with the story of her role model for aging well, can be regarded as a good example of what a "real retirement" can offer as opportunities, such as having time to follow ones' routines and having an organized life and aging well. However, preventive behavior, even if healthy, is often being performed out of fear of showing or acting out ones' real chronological age. It is not only meant to prevent illness but, also to prevent old age from happening.

Chronological Age Is Obsolete

All in all, chronological age is not believed to represent an appropriate indicator of aging anymore. Other criteria such as being active and showing interest in the world are considered essential. If an elderly person does not "*act the part,*" he or she does not fit the elderly stereotype and is placed outside "the seniors" group. Moreover, according to some participants, the whole idea of chronological age or the notion of clear life stages has disappeared.

For example, **Christoph** talked about his difficulties in guessing a person's real age (i.e., chronological age). **Christoph** owns a small café with a furniture shop together with his life partner. The café shop is situated on a busy street in one of the fashionable Berlin areas. Being at his café six days a week, **Christoph** gets to meet older people almost daily. Since the café has reasonable prices and a cosy atmosphere, it attracts elderly customers and older people from the neighborhood who use the coffee shop as their meeting point. Nevertheless, **Christoph** finds it difficult to guess the age of some customers because they do not look like the stereotypical image of the old person:

> We grew up with this cliché that old age means having wrinkles, walking with a stick, that old people are always at home and do nothing or they drink coffee all day or sit around in an institution for old people...but this cliché does not match todays reality. Being at the coffee shop all day I get to see many active seniors, sitting here with their I pads, so you cannot recognize someone is 70 just from their clothes or their looks, this has changed, so I cannot really say what the indicator would be (...) for myself I believe you are old when it becomes clear that you become another person, that more illnesses will follow....

Even if the idea of chronological age is obsolete, the concept of **change** endures. As **Christoph** explained, one "*becomes another person,*" meaning that one has transformed his or her former self into something else. Unfortunately, this "new," different person is described in terms of decline and illness showing that the negative stereotype of aging prevails, even when it remains hidden. Not looking or acting ones' real age is valued as a positive thing, while showing the real chronological age and thus embodying the negative stereotype is considered bad.

Other participants who agreed with the irrelevance of chronological age emphasized the importance of ***life experience***. The latter can make a young person think and act as someone older and thus transcend chronological age boundaries. However, negative life experience is associated with old age. For example, **Thomas** talked about knowing a girl who

> was at 30 already so old... although according to her age she was young, she had experienced so many terrible things that she was old ...

Negative thinking patterns, rigidity of thought, and lack of activity can make a young person seem older. For instance, **Alexandra** argued that

> Some people are **already ancient in their thirties**, they are so **rigid**, so obsessed with their ideas and do not want to change or accept new points of view. My ex-husband **was already old with 30, he was rigid,** had clear ideas about what is wrong and what is right, and he even looked old, he complained a lot and died relatively young. Somehow weird, he was always self-centred, never did things for other people and my daughter told me that he and his new wife only talked about illnesses and visited the doctors often and of course doctors, they always find something wrong, that is their job…

Thus, the onset of old age depends less on chronological age and more on thinking patterns and actions. Openness to new experiences, social engagement and activity embody the subjective youth and constitute the positive aging stereotype. In contrast, rigidity, illness, and decline are part of negative aging stereotypes.

In accordance with the saying "one is only as old as one feels," one may add that one is only as old as one acts. The "real" age of other people was perceived through a ***subjective lens***. Thus, a twenty year old may be perceived as "already old," whereas a seventy year old can be perceived as "still young." To further illustrate her point, Alexandra described a woman she admires. Her role model is in her seventies, but in Alexandra's view "still young" and successful in her work environment. **Alexandra**, just as the other precarious interviewees, is aware of the importance of working in old age and thus has chosen as a role model a working woman in her seventies. The indicators for old age are **subjective**, but they become real and objective in social contexts (e.g., work context) when ones' abilities are compared to the young:

> me personally I feel old when I perceive my limitations in the work context. I notice that I cannot think or act quick enough. I got to know a very interesting older lady, she was 70 when we worked together on a project. She was an IT specialist and she was still very good at her work, developing computer programs and calculating, amazing! Everybody respected her. But during the conference she sometimes would fall asleep for a few minutes. Everybody noticed, but no one said anything because they liked her a lot. She confessed to me during dinner that she used to be so fast, like she would walk fast to the metro, but now she notices how younger women would be faster than her and walk past her and get into the metro before she does. These are the moments when she is reminded she is not as young anymore. She still works, does not want to be obliged to live from her pension and she is thin and still attractive. She has her grandchildren and daughter and goes on holiday with them, she is active, but she notices that others can walk faster than her. This is for her a sign of old age. **I believe we all have our own indicators for old age**

Indicators of old age described by Alexandra comprise a decline in productivity and activity level, suggesting a rather negative image of old age. On the one hand, the old woman is described as a positive aging role model and she is still respected at work because of her competences. On the other hand, the woman herself feels old because she notices she is not "*as fast as young people are.*" This provides an example of the potential discrepancies between objective and subjective perceptions of old age and older people.

Retirement age used to be a "reliable" indicator for the start of old age. Nevertheless, since the relation between work and retirement has been reorganized (Phillipson et al. 2018; Polivka and Luo 2015), it became more difficult to define old age within a context of risk and precarity. Nowadays, entering retirement is not a linear transition from work to being dependent on social security (Vickerstaff et al. 2015), but constitutes a question of choice and options. Some people retire around the age of 65, others enter early retirement, some keep working after retirement, and others do not retire at all because they have no work to retire from. Precarious individuals are part of the latter category. Since they have no stable work contracts, they also have no clear age threshold for entering retirement. Furthermore, they cannot afford to stop working since they have no clear pension plan to rely on.

There is a link between chronological age being obsolete and the "treacherous aging body." Since aging is first noticed in the body, one risks showing the real, chronological age that is reflected in bodily signs. Thus, one needs to invest more efforts into body care and hiding the real age. Participants believed that by altering ones' appearance or way of thinking, one can transcend chronological age. In principle, this may mean more freedom since one is not stuck with the label of being of "a certain age." Although the fact that chronological age loses its relevance may sound as a positive view on aging, it is a deceitful view on old age. It actually stands for the denial of old age and the desire to stay young. It confirms the critics brought to the positive aging concept, namely that it actually promotes staying young rather than growing old gracefully (Smirnova 2012; Andrews 1999). On the whole, both the objective and subjective traits that identify a person as being "old" are negative and point to a negative view on aging in both precarious and financially secure individuals.

No Image Is a Negative Image—"Being Wrapped up in the Moment"

Some precarious individuals prefer not to think about aging at all. For example, **Christoph**, who owns a café together with his partner, is proud that he could start up a business. However, the competition is fierce, and the café shop does not bring in enough money for a secure old age. Because they do not have savings for retirement, **Christoph** and his partner qualify as precarious workers even though they own a coffee shop. Despite the financial difficulties, their start up business is still something that **Christoph** enjoys doing and would not give up. One sure thing for him is that he would like to stop working when he is older. He would prefer to volunteer in his late life, but at the moment this seems like an unrealistic dream since he and his partner will have to go on working to make a living.

> I cannot imagine that I will be still working in my seventies, maybe I am too lazy, or I am not the person that is so mentally active, but I think that the active time lies in my past, and I would like to engage maybe socially as a volunteer in old age but not be financially productive.

Christoph mentioned he had no concrete views or dreams concerning his old age. Furthermore, he did not have a clear representation about how life will look like in his old age:

> Personally, I am somehow hanging in a vacuum, I constantly reflect what to do with the shop to make it run better so that there will be some money for old age. But the present is so much on my mind that I do not get to think about the future, I am undecided, I expect all and nothing. (…) I do not have this energy to plan, I want to have my peace and quiet, with the shop I have a lot of work. I would like to enjoy more sun, eat tasty food and have good quality of life in my late years. Now, the business takes 6 days of the week away, so it is hard to plan the future.

A short-time perspective is characteristic of a negative view of aging and has negative health implications in the long run (Wurm et al. 2010). One can wonder if we always need to plan the future in order to be happy in the moment.

Similar to **Christoph**, **Michael**, a freelance language teacher in his late fifties, believes that positive aging means not to form long-term plans. Michael explained how planning for him just means having more worries about the future. He strongly believes old age will not turn out as planned anyway and therefore planning is pointless. Moreover, lack of planning is connected to enjoying life and ensuring a happy old age:

> If I think about being happy in old age, I believe I should not plan, because if I plan, then all these anxieties emerge! I ask myself what if my plans fail? I believe it is better to live in the moment, also when you age, only this can lead to happiness in old age. I do not want to think what I will do during my retirement, if I have this pension anyway because this will make me unhappy, I try to live in the moment while I am ageing. Children do this, and young people live in the moment. Of course, this is different in old age, when you live in the moment but with all your experiences, and you should not let yourself burdened by your life story and biography, but really live in the moment with all your knowledge of the world

The idea of living in the moment with all ones' knowledge of the world reminds of wisdom theories (Thomas and Kunzmann 2013). It may also reflect the dissonance that precarious individuals perceive between their own view on aging (i.e., precarious aging) and the social image of active/positive aging, embodied by the happy, healthy, productive old person. Since precarious individuals cannot identify with the latter, they either have a negative view on old age or chose not to have a view at all.

This refusal to think about aging is similar to the denial of chronological age. Denial of aging reflects a "hidden" negative view on aging. It means that people refuse to think consciously about aging, but that does not mean that they do not have views on aging that they are not aware of aging. On the contrary, these hidden negative views on aging are in a way more dangerous since they can be activated automatically (see Levy 2009), without the awareness of the individuals and thus are more difficult to change. This kind of hidden negative view on aging was typical for precarious middle-aged participants. It may have been connected to the feeling of no future since, if one does not perceive a "good future," why make the effort to think about and plan old age? If the time perspective is limited and the feeling of control over life circumstances is low, it makes sense to enjoy the moment and postpone thinking about old age. Both the short-time perspective and low control over one's life are expressions of precarity that are reflected in the negative views on aging of precarious individuals.

Positive Views on Aging Reality or Wishful Thinking?

After describing the different aspects of negative views on aging that emerge in a precarious reality, one cannot help but ask if there is some positive view of aging? And if positive views on aging do exist in a context of precarity, what are they? Nevertheless, during the interviews, participants named some positive aspects concerning aging in addition to the negative views discussed above.

Hopes related to aging comprise feelings of satisfaction, positive expectations about finding new tasks that would give meaning to old age. In accordance with these theoretical definitions, financially secure and precarious participants alike hoped to reach a feeling of well-being or satisfaction in old age. However, financially secure people mentioned more hopes for old age as compared to the interviewed precarious individuals. The first mentioned positive aspects related to their retirement, a time in their lives where they can rest and give meaning to their lives outside their regular work patterns. Precarious individuals only mentioned some positive phantasies about their old age when they were concretely asked what they associate with the term "positive aging." Since for them retirement does not really exist, daily life would not change so much in their older years. Unless they would experience bad health or other problems that would keep them from continuing their work, precarious individuals would continue to cope with the precarity of their lives.

Retirement represents a future reality for financially secure individuals. For them, life stages are still an important notion since life after retirement is regarded as Clara described a "*liberation, a time for oneself*." Even if concerning aging itself, financially secure individuals have a negative view, regarding retirement they have more positive representations. As discussed in the above sections, aging is associated with health decline, diminished attractiveness, tiredness, physical signs (e.g., gray hair, wrinkles). On the contrary, financially secure people manifest a positive view of retirement as a new life stage. Of course, financially secure individuals can afford to retire and dedicate their newly found free time to friends, family, hobbies, or social work. If retirement could occur without aging, it may represent a perfect scenario.

Hopes of an Active Old Age in Financially Secure Individuals

For financially secure individuals, hopeful thought patterns are connected to social engagement behavior and the expectation to put their life experience to good use. Engagement refers to constructive behavior, such as making concrete plans for old age and thinking about how to use the retirement time in meaningful ways. Some participants found meaning in doing something useful for others; for example, **Barbara**, a clerk approaching retirement, argued that

> one should not stick just to family, one should get involved in the public sphere, take on tasks for the general good of society

What **Barbara** expresses reflects the principle of generativity, namely of doing something for future generations (Schoklitsch and Baumann 2012). Erikson was the first to define generativity as a "concern of establishing and guiding the next generation" as part of his life stages model (Erikson 1950). He attributed generativity to the tasks of middle age. Further authors have integrated generativity as a key component of successful aging (Baltes and Baltes 1990; Kruse and Wahl 2010).

To have a task in old age is considered very important in the group of financially secure participants. For example, **Bruno**, a successful business man, describes the relevance of having "a task" in old age:

> because when you don't have your work and you do not have something to replace it with-…you feel stupid and lonely…you must figure out how to fill your days when you don't work anymore

What **Bruno** describes shows us that without a task in old age, the gained years of life experience are quite useless. One needs to be productive in order to feel useful in old age. Once again, a negative view of aging is hidden behind the positive discourse.

Overall, knowing that one has a secure pension plan makes old age look more promising. It stimulates one to phantasies what one will do with the free time during retirement. As **Clara**, a school teacher in her early sixties, mentioned: "*For the first time in my life, I will be able to see myself, to attend to my needs.*"

Nevertheless, it is of key importance that these pastimes are chosen individually and not dictated from outside. Some financially secure individuals clearly expressed the fact that they do not identify with existing positive images of aging such as "*grandma wit rosy cheeks*" (Charlotte, journalist in her mid fifties) or "*elders fishing in a pond*" (Britta, a bank clerk in her early fifties). Even if they are not quite sure about what they will do in old age, the relevant thing is to choose themselves. In this sense, the positive view of aging comprises the active aging philosophy where people are autonomous and "design their own old age." All in all, what financially secure individuals describe is the view that is projected by active aging policy. One needs to be productive and useful for society in old age in order to be considered a successful ager.

Hopes of Wisdom and Life Satisfaction in Precarious Individuals

Precarious individuals' positive representations of aging relate to old age being an achievement it itself as **Daniel**, the freelance music teacher, explained:

> it is a great chance to become old as many people do not reach old age because of illness or so…

Old age is not considered to be entirely bad. For instance, the acquired life experience allows one to be more relaxed about life's' problems and challenges, **as Kai** suggested "*I would say it gets easier after you turn 40*". Thus, even if problems may increase with age, at least the ability to deal with difficulties may improve.

Wisdom embodies the ability of handling difficulties effectively, a quality that is acquired with aging and it is associated with well-being. Emotional maturity helps to deal better with problems. For example, **Daniel** expressed his hopes about his own development of wisdom in old age:

> I hope I will become more satisfied with life, with other people and with myself as I grow older

All in all, participants who are employed in precarious jobs tried to figure out a way to make ends meet in old age and develop the necessary wisdom to deal with life's problems in a relaxed manner. Similar to financially secure individuals, precarious individuals oppose images of aging that are projected from the exterior and try to design their own positive alternative views on aging. For example, **Michael**, talked about his ideal self in old age:

> My image of a positive aging is that I will take a walk with a nice walking stick maybe but definitely not a waking aid, and I will go to the evening concert in the Philharmonic and there I will be surrounded by young people. For me taking part in cultural life would be very important, if I could not take part in cultural events, I would be very sad, but I could of course see concerts also online. At the moment, I cannot really picture this, because I use internet only for pragmatic things, for work, so I am a bit suspicious about it, but with time I may need it for other things, something like age tailored internet…virtual space will become more important…

Wisdom implies self-acceptance, and this may be a very important positive aging views for precarious individuals. Furthermore, self-acceptance is a trait that seems to lack in the accounts of the financially secure individuals since they do not seem to accept their aging process.

Wisdom is associated in the literature with a good life (Yang 2013) and is associated with internal emotional and cognitive development. Wisdom was also defined as success in handling problems (Baltes and Staudinger 1993). Thus, wisdom can be considered part of a positive view on aging. Wisdom represents a trait that a person acquired through personal effort and therefore is in line with the idea of successfully achieving something in ones' life. Wisdom represents the reflection of ones' efforts to lead a good life. It refers to mastering challenges during a longer period of time (Yang 2008, 2011, 2012). Furthermore, it implies that one can be active also by contemplation and reflection on life and ones' experiences. It does not imply that one should be productive in order to be happy or socially integrated. In this sense, wisdom in old age represents an alternative positive view on aging that precarious individuals can identify with.

It is not surprising that precarious individuals talk about wisdom since this is often related to hardship or finding happiness despite hardship (Yang 2013). Wisdom is also related to generativity, since it is often described in the literature as reaching success in ones' life by helping others (Kunzmann 2004; Yang 2008, 2012). A good life in respect to wisdom is defined as a life where one is satisfied with oneself and offers happiness to others (Yang 2013; Bluck and Glück 2004; King et al. 2004). The road to wisdom is seen as one that implies hardship or even pain and thus fits well with the context of precarity. The latter may even contribute to achieving wisdom. Hardship motivates people to search for meaning and goodness in life's problems (King et al. 2004). In similar ways, precarious individuals may search for meaning and positive aspects of uncertainty and insecurity. Hardship and pain can be antecedents of wisdom (Yang 2013). Studies have also shown that wisdom has a great association with middle age (Staudinger 1999; Yang 2008). While for generativity one often also needs material resources, for achieving wisdom this are not necessary. This may be one reason why financially secure individuals focus on generativity, while precarious individuals concentrate on wisdom.

References

Andrews, M. (1999). The seductiveness of agelessness. *Ageing & Society, 19*(3), 301–318. https://doi.org/10.1017/s0144686x99007369.

Baltes, P. B., & Baltes, M. M. (Eds.). (1990). *Successful aging*. Cambridge University Press. https://doi.org/10.1017/cbo9780511665684.

Baltes, P. B., & Staudinger, U. M. (1993). The search for a psychology of wisdom. *Current Directions in Psychological Science, 2*(3), 75–81. https://doi.org/10.1111/1467-8721.ep10770914.

Bluck, S., & Glück, J. (2004). Making things better and learning a lesson: Experiencing wisdom across the lifespan. *Journal of Personality, 72,* 543–572.

Erikson, E. H. (1950). *Childhood and society*. NewYork: W.W. Norton and Company Inc.

Fung, H. H. (2013). Aging in Culture. *The Gerontologist, 53*(3), 369–377. https://doi.org/10.1093/geront/gnt024.

King, L. A., Eells, J. E., & Burton, C. M. (2004). The good life, broadly and narrowly considered. In P. A. Linley & S. Joseph (Eds.), *Positive psychology in practice* (pp. 35–52). Hoboken, NJ: Wiley.

Kruse, A., & Wahl, H.-W. (2010). *Zukunft Altern: Individuelle und gesellschaftliche Weichenstellungen* [The future of aging: Individual and societal directions]. Heidelberg: Spektrum. https://doi.org/10.1007/978-3-8274-2200-2.

Kunzmann, U. (2004). Approaches to a good life: the emotional motivational side to wisdom. In P. A. Linley & S. Joseph (Eds.), *Positive psychology in practice* (pp. 504–517). Hoboken, NJ: Wiley.

Levy, B. (2009). Stereotype Embodiment. *Current Directions in Psychological Science, 18*(6), 332–336. https://doi.org/10.1111/j.1467-8721.2009.01662.x.

Phillipson, C., Shepherd, S., Robinson, M., & Vickerstaff, S. (2018). Uncertain Futures: Organisational Influences on the Transition from Work to Retirement. *Social Policy and Society*, 1–16. https://doi.org/10.1017/s1474746418000180.

Polivka, L., & Luo, B. (2015). The neoliberal political economy and erosion of retirement security. *The Gerontologist, 55*(2), 183–190. https://doi.org/10.1093/geront/gnv006.

Smirnova, M. H. (2012). A will to youth: The woman's anti-aging elixir. *Social science and medicine, 75*(7), 1236–1243. https://doi.org/10.1016/j.socscimed.2012.02.061.

Schoklitsch, A., & Baumann, U. (2012). Generativity and aging: A promising future research topic? *Journal of Aging Studies, 26*(3), 262–272. https://doi.org/10.1016/j.jaging.2012.01.002.

Staudinger, U. M. (1999). Older and wiser? Integrating results on the relationship between age and wisdom-related performance. *International Journal of Behavioral Development, 23,* 641–664.
Thomas, S., & Kunzmann, U. (2013). Age differences in wisdom-related knowledge: Does the age relevance of the task matter? *Journals of Gerontology. Series B, Psychological Sciences and Social Sciences, 69*(6), 897–905. https://doi.org/10.1093/geronb/gbt076.
Vickerstaff, S., Phillipson, C., & Loretto, W. (2015). Training and development: The missing part of the extending working life agenda. *Public Policy and Aging Report, 25*(4), 139–142.
Wurm, S., Tomasik, M. J., & Tesch-Römer, C. (2010). On the importance of a positive view on ageing for physical exercise among middle-aged and older adults: Cross-sectional and longitudinal findings. *Psychology and Health, 25*(1), 25–42. https://doi.org/10.1080/08870440802311314.
Wurm, S., & Huxhold, O. (2012). Sozialer Wandel und individuelle Entwicklung von Altersbildern [Social change and individual devolvement of views on aging]. In F. Berner et al. (Eds.) *Individuelle und Kulturelle Altersbilder* [Individual and cultural views on aging], 27–69. https://doi.org/10.1007/978-3-531-93286-6_2.
Wurm, S., Warner, L. M., Ziegelmann, J. P., Wolff, J. K., & Schüz, B. (2013). How do negative self-perceptions of aging become a self-fulfilling prophecy? *Psychology and Aging, 28*(4), 1088–1097. https://doi.org/10.1037/a0032845.
Wolff, J. K., Beyer, A.-K., Wurm, S., Nowossadeck, S., & Wiest, M. (2018). Regional impact of population aging on changes in individual self-perceptions of aging: Findings from the German ageing survey. *The Gerontologist, 58,* 47–56. https://doi.org/10.1093/geront/gnx127.
Yang, S. (2013). Wisdom and good lives: A process perspective. *New Ideas in Psychology, 31*(3), 194–201. https://doi.org/10.1016/j.newideapsych.2013.03.001.
Yang, S.-Y. (2008). A process view of wisdom. *Journal of Adult Development, 15*(2), 62–75. https://doi.org/10.1007/s10804-008-9037-8.
Yang, S.-Y. (2011). Wisdom displayed through leadership: Exploring leadership-related wisdom. *The Leadership Quarterly, 22,* 616–632. https://doi.org/10.1016/j.leaqua.2011.05.004.
Yang, S.-Y. (2012). From personal striving to positive influences: Exploring wisdom in real-life contexts. In M. Ferrari & N. Weststrate (Eds.), *The scientific study of personal wisdom* (pp. 115–135). New York: Springer.

Chapter 7
Preparing for Old Age

Introduction

Apart from asking about the meaning of old age, a decisive question is how do middle-aged people with different social backgrounds actually prepare for achieving a good old age? Previous literature states that dreaming about successful aging or just wanting a wholesome old age is not enough in order to achieve this goal. According to health policy guidelines (Lassen and Moreira 2014; Walker 2009), active aging needs careful preparation. Just as we saw in Chap. 6, views on aging differ between financially secure and precarious individuals. Thus, one may presume that preparatory efforts differ between those who have a secure pension plan and those who do not. Precarity can be thought to influence preparations for old age in several ways. For example, precarity may establish certain barriers that could hinder preparatory efforts. In the present chapter, I explore the concrete preparations that middle-aged individuals undertake toward their old age and compare them to the preparations that financially secure individuals carry out.

During the interviews, middle-aged precarious and financially secure people were asked how they prepare for old age and what helps (i.e., facilitators) or blocks (i.e., obstacles) them when making arrangements for old age. Findings from the study showed that preparations for old age were quite different for precarious and financially secure individuals. The first part of the chapter presents the results concerning preparations of precarious individuals, while the second part of the chapter is dedicated to the preparatory efforts of their financially secure counterparts. Barriers and resources that precarious and financially secure individuals consider relevant for achieving a positive old age are also discussed.

I. C. Crăciun, *Positive Aging and Precarity*, International Perspectives on Aging 21, https://doi.org/10.1007/978-3-030-14255-1_7

To Prepare or not to Prepare for Old Age?

Even if precarious middle-aged Germans are preoccupied by the idea of aging and its implications (see previous chapters), most of them have not formulated any concrete plans to ensure a good old age. In fact, almost all interviewees with precarious jobs mentioned that their main preparatory strategy is not to plan at all. In their view, planning for old age requires several resources (e.g., time, energy, financial assets, or secure pension plan) that they do not possess. Thus, precarious people usually decided not to plan for old age and "*live in the moment.*" Mostly the dilemma of precarious individuals can be understood as "to prepare or not prepare for old age" and answering the question whether it is relevant to prepare something that is anyway uncertain. In the following, I will explore the strategies that precarious individuals considered important for preparing for old age in times of precarity. These include the following: thinking short-term and focusing on the present, staying optimistic in face of adversities, developing work skills, investing in health, building social networks, and using the family as a social and emotional resource.

Thinking Short-Term

Individuals with precarious backgrounds emphasize the importance of **not planning** their old age as doing this might generate anxieties. Since often precarious individuals do not picture a future for themselves, they prefer to enjoy the present and avoid thinking about how their old age will look like. In their view, one cannot control the future and trying to do so, only causes further worries. Because precarious individuals perceived a brief time perspective, participants with precarious backgrounds tended to focus on short-term projects. For example, they would plan to apply for a new job, learn a new skill that would make them more eligible on the job market, or just organize some leisure time activities. This lends support to previous findings that show that goal selection depends on the perception of time (Carstensen et al. 2003; Fung and Carstensen 2006). According to Carstensen et al. (2003), people choose between expansive goals (i.e., acquiring information) or emotional goals (i.e., balancing emotional states) depending on the time they perceive they have left to live. When time is perceived as open-ended such as in younger years, expansive goals are usually pursued (e.g., making new friends, learning new skills by participating in varied activities). When the time left to live is perceived as limited, emotional goals are pursued. This happens because people need emotional support in the present and not for some time in an uncertain future. Precarity is associated with a limited time perspective and thus should inspire people to choose emotional goals. However, since precarious individuals need to make social contacts for work, they invest in both expansive goals and emotional goals. The expansive goals have a strong focus on the present (e.g., finding work), while emotional goals help them both for the current emotional regulation and for ensuring social support in later years.

Even if the precarious situations can be very different, the reasons for not planning are very similar among the precarious individuals. Participants complained that they do not have time to sit down and reflect on their future. Also, some mentioned that they started thinking quite late in life about pension prospects and they are overwhelmed by the lack of present security in their lives. For instance, **Christoph**, the owner of the furniture café in his late fifties, summarized the reasons why he chose not to make specific plans for his old age:

> For me personally it is like this that the shop brings in little money so when we add it all up, in the end we are left with very little, so we cannot look forward to our old age, there is no financial solution for our old age at the moment, I do not know how we will organize ourselves in old age.... I have no idea, from the beginning we did not have a great start like a pension insurance, we did not pay it and we also thought it is all right, we have not really thought about old age. And now I do not know what is right. I try to organize some insurance at the moment, but I believe this will also not guarantee 100% security because something can always go wrong, and then I saved all the money for nothing! Also, I am not that ambitious, I find it difficult to motivate myself.... Advertisements encourage us to save for old age, and of course this is a good business for them, and maybe they are also right, but old age is not really secure anyway!

For participants like **Christoph**, the prospect of old age is precarious in itself, and thus, saving money or making concrete plans seems pointless. The awareness of precarity generates feelings of depression and lack of motivation to come up with solutions. Furthermore, it offers possibilities to rationalize ones' decision not to plan or save for old age since "*the outcomes are uncertain anyway.*"

Similar to the idea of having started too late to plan and save for old age, some participants like **Kai** (freelancer working in academia, in his early fifties) expressed their regret of not having started to plan earlier:

> it would be nice if you could already begin as a child to think about it [old age], but now, unfortunately you cannot plan it [old age]...

For other participants, precarious work just takes up too much of their time and energy and they have little left for self-reflection. For example, for **Corinna**, being a freelancer means she has too many things to do on a daily basis and thus she has little time for herself. This means that she has little time to think about her future goals. For **Corinna**, "*living in the moment*" represented an effective coping strategy for dealing with the uncertainty of her work reality. She considers planning would only increase her anxieties about the uncertain future. Thus, **Corinna** preferred to focus on her present uncertainties and solve each problem as it comes:

> I know that people are expected to think about what they will do when they retire, like travelling, but I think I will not afford to do that financially anyway so I do not think about what I will do with 65 or 70, I am much too busy thinking how to survive in the moment...I only wish to be healthy, but what exactly I will be doing I do not know, I have no dreams about old age, **I dream about the here and now**, not about what I will do when I will not work anymore, because I do not think that will be the case (,,,) actually what helps me to be more relaxed and not get into a panic is exactly not to think about the future, it is actually one of my most important resources...

Just like **Corinna**, several other precarious participants considered "*not thinking about old age*" a resource for positive aging. This may sound like a paradox, but in the context of precarity, it can prove to be an effective coping mechanism. In this view, short-term plans, if successful add up to positive aging, since old age is not seen as an end state but a continuous process.

The precarious living situation with its associated feeling of "no future" (Foti 2017), obliges individuals to focus on the present and postpone thinking about the future. Perceiving a short-time frame makes people avoid planning for the future. Plans may work well in a secure environment. Nevertheless, surviving the insecurities of the present take up too much energy and time. In an insecure context, characteristic for precarity, people may not make plans because they do not believe that they can enact those plans. Coping plans (i.e., thinking about obstacles and ways to overcome these) may be more helpful in such an insecure life setting.

Fostering Positive Thinking

Precarious individuals often apply positive reframing of their precarious life instead of planning as a strategy for dealing with stress and preparing for the future. For example, **Christoph** mentioned he uses positive reframing in the sense that he changes his expectations concerning the future or focuses on the positive aspects of his present reality:

> I expect rather positive things than just focus on the negative aspects and that is why I do not think so much about ageing… I do not plan…it is also a job in itself to plan and I do not want to…

Despite having mostly, a negative view on aging, as we have seen in previous chapters, precarious individuals recognize the importance of trying to stay optimistic regarding the future. Most of the precarious participants recounted using positive thinking as a defense mechanism against daily worries. Positive thinking helps precarious individuals cope with the uncertainty of everyday life. For example, **Kai** (freelancer working in academia, in his early fifties) explained how his precarious existence forces him to constantly search for solutions:

> you can't lay down your weapons so to speak…negative thinking takes the energy away, energy that one needs in order to survive on a daily basis (…) If I had an insurance… this would be like a psychological resource, it will give me some security, I could say I can look into the future with trust…So better think positively, it might help one day…

Positive thinking as a cognitive resource may compensate for lacking financial security to prepare for old age. Another precarious middle-aged participant **Mamadou** (a taxi driver in his early fifties, originally from an African country, but living in Germany for more than 20 years) talked about the importance of the general attitude toward life. In his view, one needs to have a positive attitude in order to cope with the stress of the precarity experienced in everyday life. **Mamadou** supports a proactive coping strategy concerning problems that occur in everyday life:

> I do not know what happens in the next two or three years, but I do not think about this, I prefer to think there will always be a solution to all my problems. Life will go on. In life there are always problems, some problems we make ourselves, others just come upon us, but there is also a solution to the problem if you want to solve the problem. But sometimes people just run away from their problems. The problems will not go away if you avoid them, you have to solve them.

Investing in Health

Despite not actively planning for old age, some precautions are taken by precarious individuals especially in what concerns crucial resources for an active old age such as health. For instance, all interviewees mentioned going for regular medical check-ups. In this context, "*being informed*" about ones' rights as a patient was considered of key importance. When one does not have a secure job, the probability that one has a problematic health insurance is high. Thus, it is crucial to find an efficient doctor or to discover affordable solutions to keep healthy since health is relevant for work.

For instance, **Alexandra** (freelance interpreter in her early sixties, no secure pension plan) talked about how she likes to keep well informed. She is not just interested about health issues in general, but concretely on how to "***defend herself***" when doctors propose treatments that are too expensive. The negotiations with the health system and its representatives are focused around money and time issues:

> I go to medical check-ups, not excessively, but you know the basic stuff. I do not go to a family doctor since they always want to check every possible thing, but I go to an experienced internist who looks only at the relevant things. He does not try to find something, he takes my complaints into account (…) I also go to an optician, to a gynaecologist, one has to, no? (…) but I do not want to make a hobby out of going to the doctor (laughs), some people go to the doctor just to have someone to talk to, this is not my case since I am busy working. My family doctor is very nice, but she has no discipline! I always must wait hours at her practice! There are many old people in her waiting room talking about their illnesses! And she listens to every patient, so consultations take so much time! I do not have time, so I do not go (…) it is good to be your own boss! I must take care of myself in general, so why not my health? I managed to defend myself against my dentist who wanted to do a preventive treatment that was too expensive for me. I thought I was just working too much and was tired so of course this would influence the health of my gums. I told him I need a holiday, not a second prophylactic treatment for my gums (…) you must be informed, you have to think about your health actively, you have to be able to defend yourself against unnecessary treatments!

Not just holidays, but also work itself can represent a pleasant way of keeping active and ensuring ones' mental equilibrium especially if one enjoys what they are doing. For example, **Carmen**, a Latin American woman in her early fifties, works as a part-time architect. Besides bringing up her two children, she enjoys her work very much even though she feels she has two full-time jobs. Parenting represents an around-the-clock activity and her part-time job does not really allow her paid holidays. Thus, she has to decide for herself when to take a break. In order to help her with this decision, **Carmen** says she tries to "listen to her body." This means that when she feels tired, she takes some time off for herself. All in all, she is happy she has found even a part-time job in an architecture office where she also enjoys working. Having two children to take care of, she does not want to work full time. Also, full-time jobs in her field are very competitive. When she first came to Germany and before she had children, she worked full time and supported her husband financially. However, after she became a mother it was her husband's turn to follow his career path. Now, they are divorced, and **Carmen** has full custody of her children. Since she is in her early fifties, when she applies for jobs, she can only find part-time positions that allow her the flexible schedule she needs.

> my work is fun, and then if I can earn some money with it, it is even better! Work is one way to keep healthy, and of course I am careful with my diet, to eat healthy, so work and healthy eating are about the only things that I can afford to do now in preparation for old age…

For **Carmen** as for other study participants with precarious jobs, work is associated with being healthy and active. Automatically, health behaviors are linked to sustaining the energy for work. Prevention activities become part of daily life.

Acquiring Work Skills for Later Life

According to the interviewed precarious individuals, they do not have to worry about a meaningless old age. Work and constantly looking for new jobs always gives them something to think about and keeps them constantly active. It may not be a very positive meaning for old age, but it surely provides some meaning to old age. Moreover, some of the participants really enjoy their work even if it is precarious (e.g., short-term or badly paid). Thus, in order to deal with their concern that they may not be able to continue working in old age, they develop strategies to continue working.

The preparatory actions they usually undertake are mostly directly linked with their jobs and the plan to keep on working as long as possible. Precarious individuals must stay mentally and physically fit, to keep the jobs or get new jobs in their older years. Several participants talked about their strategies concerning being active on the job market. These strategies included staying informed concerning job opportunities, keeping up to date in their fields and building social networks at work in order to make sure they will continue to be hired.

For example, **Daniel**, being a music teacher and musician finds collecting partitures and music programs important for his future work and for his continuous self-development. He recounted how every time he goes to a concert, he takes the program with him and puts it aside for later. One day when he is older, he will have time to practice these music partitures. Apart from this, he makes sure he keeps his options open in several directions.

> For myself I am sure I will not be able to live from my pension, so I will have to work. I can imagine myself continuing to teach. Probably I will do that less than now, and with maybe only tutorials and not music theory classes. I also am a musician and give concerts, so I could

> continue to do that. For example, I play the organ in a church, so I could do that. …I can imagine when I am older it will be more difficult to give concerts for a large audience but teaching an instrument on a tutorial basis I will be able to handle. Probably it will be more private than public, but it all depends on the energy and health I will have. At the moment, the music school provides me with pupils. This may become more difficult when I get older. I keep in contact with many schools. I make sure that I have other advertisement possibilities for my teaching such as web sites or just word-of-mouth recommendations will work just fine when I am older.

For such participants like **Daniel**, even if they do not have a concrete agenda for old age, they invest in their personal resources (e.g., skills) and think about working alternatives for their later years.

Developing Social Networks

Social networks are not just beneficial for strengthening work contacts and opportunities but serve to provide a sense of safety in everyday life. ***Precarious individuals*** feel they need to be good social networkers. Social contacts are vital for finding work but also for the future perspective of living together or helping each other in their old age. In other words, when financial resources are lacking, social resources may compensate.

For example, **Corinna** (freelancer in her mid-fifties, no secure pension plan) talked about her intention to share a flat with friends when she will be older. In this context, she mentioned her prior positive experiences with multigenerational forms of living when she shared a flat with friends who had children:

> I think I will live with my friends when I am old, we will share a flat, we can cook together, it is cheaper, we can do fun things together….

In addition to being important for sharing living expenses in old age, building social networks for the future was considered a relevant asset for remaining socially engaged.

Nevertheless, not all interviewees were so happy about the idea with sharing their flat with other people in their old age. For instance, **Michael** (freelance language teacher in his late fifties) described himself as a "lone wolf." For him, being independent contradicts the need to rely on others for help when material resources become limited. However, he realizes that social networks may turn out to be progressively important with age:

> I can only imagine it is horrible to depend on someone else! But when it does not function otherwise, what can one do? It could be a beautiful discovery in old age! It may be fun, I am travelling through life as a loner, but I have the feeling I find something beautiful everywhere and I am not alone even if I am not sociable. But it can happen that all will change in old age, that I will want to spend time with other persons, that I would take part in activities organised in groups such as a city tour or walking groups or a card playing group perhaps…

An interesting point concerns how precarious individuals imagine the idea of partnership and living arrangements in old age. For example, **Michael** prefers to live alone, even if he has a stable girlfriend. However, Michael realizes he might have to change his ideas when he is older and may need to adapt to living with others or at least with his life partner. Even if romantic partnerships are regarded as a factor for security in old age, some precarious individuals such as **Michael** need to negotiate between their desire for independence and the desire to be in a couple in old age.

> I would like to stay in my own apartment, but there would be barriers since I live on the third floor and there is no elevator, so it will be a problem, I would need an apartment that I can reach easily for instance in a house with a lift. I live alone, and I always lived alone except for one year in a shared flat during my university time. I would like to continue living alone. I have a girlfriend, but we live in different apartments. Even if I imagine we will be together forever, we live separately but maybe we will have to change that one day. It is weird, but I guess it makes sense to live with others in old age. So, when problems come you are not alone…I can imagine myself in a small house though, with a garden ….

Daniel, the freelance musician, is divorced and lives alone. Even though he is accustomed to being on his own, he acknowledges that for him partnership and family are relevant ingredients of positive aging. Moreover, he considered social networks relevant for his own survival in old age:

> I cannot imagine an old age without a partner, it would be terrible to be alone in my old age, I cannot imagine it, it is such a dangerous thing for me and I will do anything to prevent it…when I was younger there was the topic of having children, but now I have one and she will be out of the parental home soon, so I need other aims, to be with a partner and enjoy things together, to enjoy life together, even if there are fights or misunderstandings, these also keep people together, it is weird but true (….) if you live together with other people or you have regular meetings, than you feel more secure in old age.

For those who are not currently in a relationship, there is still hope that one will "find a partner" in the future. In this respect, the interviewed precarious individuals were optimistic. For example, **Alexandra**, the freelance interpreter, explained she is confident that one can still find love in old age. **Alexandra** recounted the positive example of her mother who remarried in her sixties. Nevertheless, when reflecting on her own precarious life circumstances, **Alexandra** considers that her stressful lifestyle constitutes a barrier toward finding a partner:

> Now that the work is getting less, I have time to look around, because when I am stressed my senses are blind, I do not see men around me. So, let's see, my mother met my stepfather when she was 63, she was already a widow for a few years. For them it was love at first sight, a big passion, she was 65 when they married, so I still have a chance. It was real love, the two of them belonged together, better than my mother and my father. She has an enjoyable time, my stepfather was a bit older than her, so she had to look after him in the end, that was sad. But he had a good pension and he also took care of the financial things, so all in all she had a good time. If I have good genes than I also have a chance because I still have some time to live. I read an article about partner search on internet. That is not for me. Some people can enter relationships in a very pragmatic way, that is not my thing. And I do not want to live with somebody again. I want to stay independent.

For precarious individuals who are used to being alone, freelancers in their work, this attitude may extend to their private life. Thus, even though they want a romantic relationship, they also want to keep their independence in certain aspects such as housing or interests.

Using Family Structures as a Resource for Old Age

For the precarious individuals with an immigration background, there is still a "plan B" for old age. Namely they have the possibility to go back to their countries of origin where they have the social networks of their families. For example, **Carmen**, (freelance architect, divorced and raises two children by herself) talked about the importance of family and friends and how relevant it is to try to keep your loved ones close to you:

> I have two great children, they love me, I have a wonderful family who support me, they make sure I am fine wherever I am, I have good friends who support us, I cannot complain. I think real friends and family you cannot lose, they will be there for you, they are like an insurance in old age that I will not be alone. The family will be there for me. This is very important for me, I cannot imagine being alone. With friends, we joke that we will share a flat in old age, just 3 or 4 friends. We know each other very well and we can live together we could still do things together when we are 65, like go to concerts, theatre, opera. Not in couples, just us friends, have fun, and if one does not have money, then the others can help. Just like in a family, that I can imagine. The only thing is where we will live, here or in [Latin American country]. I would prefer [Latin American country] since there I have a nice house where we all fit. But these are only thoughts. Perhaps until then I find another love, will have a new life partner. But if that does not happen, I will have my friends.

Besides the precariousness of the work conditions, immigrants feel vulnerable because they must adapt to the cultural habits of the countries that adopted them. **Carmen** went on to explain why the contact with her friends from her own country has helped her survive in a foreign country:

> Without my friends, I would not have survived 15 years in Germany. Sometimes to meet here (Germany) depends on the weather, so you have to make an effort to maintain social relations, you have telephone, skype, every possibility these days to keep contacts. I do not want to isolate myself, I have friends in many foreign countries and I keep in touch with them.

Facing the perspective of growing old in a foreign country, immigrants reflect on the cultural norms concerning aging and old people in the country where they will age. **Carmen** mentioned some things that she does not understand about her adoptive society:

> I notice that here in Germany many older people are alone, society is different here and elderly are put in the old peoples' home. This would be unacceptable for me, for myself I think old people belong with the family, they are very useful, they can help with grandchildren. Here they are visited by family, but they do not belong in the everyday life. I don't like that. I hope to belong to the everyday life of my family for as long as possible. If not, I better die than end up in the old peoples' home. Even if that would be like a 5-star hotel, you are still alone in your room. You are surrounded by old people, and the family is the best present…

Family and community were found also in other studies with Latin American communities to be important (Hilton et al. 2012; Gutheil and Heyman 2006).

Mamadou comes from an African country and in living since more than 20 years in Germany. Although he does mostly mini jobs, he manages to save some money to send to his family in his homeland. He also sends money regularly for building a house in his home village. This house that he builds in his home country will be his insurance for old age

> If I stay here in Germany in old age, I will have to pay rent. But if I return to my home country, I will have a house waiting for me. (...) here in Europe you are left alone when you are old, but in my country, there is always someone in the family ready to take care of you, you do not need to worry, for instance there is no institution where you put your old mother or father, there in my homeland you have family and they care for you not because it is their duty, but because they want to. Here in Europe old people are so alone!

German middle-aged individuals with precarious living circumstances also consider family an important resource for old age and in general. For example, **Thomas** talked about the significant role family plays in his life. Unfortunately, his relatives live far away, in a totally different German regions and he cannot visit often because of lack of financial means not lacking time or will to see them:

> It is very important to be socially integrated, you do not necessarily have to be integrated inside the family, but also friends. Although I do believe family to be important, despite all the postmodern debates, I think family is still vital, we are human beings, we cannot just brake away from our families, we are not virtual beings, we are social beings and we are socialised inside our families. We cannot abandon this like a coat we do not wear anymore, it marked us for life. I am very close to my family, although I cannot see them often, it is so far away *small town name* and with my financial situation I cannot afford to travel there as often as I would like to.

To develop and maintain social relations, time is a key resource. However, time is a resource that precarious individuals perceive they do not own, as **Thomas** said:

> Of course, you need time, in our society we are always under pressure, in our modern or postmodern society we never have time and to have time is a great resource, I believe for positive aging it is important to have time, to build social relations...

In the following, I will explore the barriers toward a good old age, as perceived by the interviewed precarious German middle-aged individuals.

Barriers Toward a Positive Old Age Experienced by Precarious Individuals

When asked why they do not actively plan their old age, precarious middle-aged individuals mentioned several obstacles that stop them from planning old age. Such hindrances range from financial problems to cognitive barriers (e.g., lack of perceived control).

The Feeling of Lack of Control

Living in precarious circumstances means one perceives an acute lack of control over one's own time. The daily routine is structured by work or job search activities.

Instead of clear delimitations between work and leisure, the first takes up most of the day and often no time remains for the latter. Planning requires a certain structure that allows one to distribute time for different activities. When such a structure is lacking, planning becomes very difficult or even irrelevant.

For example, **Kai** talked about the difficulties encountered when preparing for his old age. Like many other members of his generation, **Kai** is aware that a healthy body is not a given, but a result of investing time, energy, and money for its maintenance. A healthy lifestyle would imply adopting a healthy diet and engaging in physical exercise at least three times a week. **Kai** works as a freelance researcher, and his earnings depend on how many projects he has. Thus, he believes he cannot afford to take breaks from work, or in his own words "*to put down his writing pen.*" For new job applications, he is required to demonstrate he has published many articles, had numerous conference presentations or successful grant applications. For his current temporary position, he needs to write research grants in order to extend his work contract. Like many in his situation, **Kai** would like to practice more sports, but faces the problem of finding the time and energy for it.

> Unfortunately, what can I say, I should practice more sport, pay more attention to my health, but in my line of work it does not work that you give only 50% of your time. I must always look for new jobs, send in applications, because for the next job they will ask me to have 8 conference presentations pro year and more than 3 articles published. This is the disadvantage, it is difficult to have it all…

Kai feels the pressure to be constantly productive, and it leaves him little time for other activities. Continuous stress of working or not being able to stop working after retirement age, gets precarious individuals distracted from health goals as Kai recounted. This is a big mental barrier toward thinking about health. For once it does not allow one to fit time for health behavior. Second, it leads to a very stressful life where one does not take breaks, holidays, and does not get enough sleep. Being constantly reminded that health is an important resource, and one needs to be healthy in order to work in old age can come as an additional stressor for many precarious individuals. It may even trigger the counter reaction for individuals faced with pressure to be healthy may engage in risk behaviors to cope with the stress of health.

However, even if someone has time in theory, sometimes the motivation is lacking, or the financial means are scarce. Contrasting to **Kai's** situation, **Thomas** (freelancer in his early fifties) admitted he could go jogging anytime because he is currently unemployed. Thus, in theory he would have plenty of time for sports, but in practice he lives a sedentary life. **Thomas** compared himself to his friend who runs 4 km every day and to himself as a young adult and used to be a good athlete since this was also a condition for his job. His previous secure job required him to have a healthy and fit body. Nowadays, as he is a freelancer after long time being unemployed, he feels unmotivated in general:

> I used to run the 500 m race, I was top fit in my thirties, but then I stopped, my girlfriend tried to motivate me to go jogging, but I can't, I am maybe lazy, I know I should do more, but with the insecurity of the financial future, of the physical future it is difficult, all people know this, but too little is being done…

Sometimes it is easier to introduce a healthy routine, if one already has a fixed schedule provided by set work hours. Without clear external time structures, too much responsibility is left for the individual to take care of health, of finding work and meaning to his or her life. When barriers are more clearly delimitated, it is easier to deal with them by formulating action and coping planning as many studies in health psychology suggest (Wolff et al. 2016; Carraro and Gaudreau 2013; Sheeran et al. 2006). For example, if one does not go jogging because the weather is cold and rainy, alternative physical exercises can be found or appropriate clothing can be used for jogging. However, when people have a general feeling of lack of control over their time, or a learned helplessness acquired during the unemployment years, it is more difficult to develop action and coping plans for physical exercise. Authors of goal theory may argue that precarious individuals should place health as a more important goal in their goal hierarchies (Gebhardt 2006). But is this not easier said than done? More work needs to be invested into considering how might one do health counseling with such persons to make health goals happen.

The Commodified Healthy Body

Thomas (started freelancing after being long time unemployed, early fifties) raised another crucial point concerning barriers to health for precarious individuals. He said he is aware of the necessity to eat healthy and he would like to pay more attention to what he eats. Nevertheless, he mentioned that "*Healthy food is expensive, you have to afford to eat healthy and that is difficult for people like me who are unem*ployed…." In many contexts, unhealthy food is less expensive than healthy products. When shopping at the super market, one would choose for instance pasta over fruit and vegetables, even when being aware of the unhealthy choices they are making. Intervention ideas should be showing "how" to buy healthy and cheap ingredients to cook food that is nourishing for the organism.

Health insurance and old age care are expensive for someone who does not have regular financial means. For example, **Kai**, the self-employed academic, has his wife and two kids on his health insurance. Thus, he mentioned:

> at the moment I have to figure out how to pay my health insurance regularly, so something like care insurance for old age or pension insurance does not come in question. I have to consider the health insurance for my children and for me something like a private health insurance that would pay for extra healthcare is unfortunately illusory at the moment.

Health represents a fundamental resource for precarious individuals. Without health, they would not be able to work in their old age and pay their bills. However, seeing the lasting benefits of health behavior requires long-term thinking. As Standing (2011) was mentioning in his book, the precarious mind is characterized by the loss of long-term thinking. Also, a long-term perspective is a characteristic of positive views of aging. As I pointed out in Chap. 6, precarious individuals hold rather negative images of aging, and focus on living in the present, rather than investing in their future. They hope that by living well in the present, they automatically ensure a good life for the future.

Investing in health requires planning. One must plan when to go jogging or swimming. Healthy meals require planning. One needs to shop for the healthy ingredients and find time to cook. In this sense, precarious individuals are more like young people, living in the moment and concentrating on short-term benefits. Precarious life obliges people to plan for surviving in the present and not for the uncertain future. Since old age is associated with fears, planning triggers sometimes the opposite reaction of generating anxiety.

For example, **Michael**, the freelance language teacher, explained the importance of not planning for old age because it generates anxiety:

> I do not want to plan my old age… in life one always fears losing something, I believe one should lose this fear in old age, one has nothing to lose anymore and if you plan all, then you have something to lose….

Since they are used to insecurity in their lives, planning to have security seems pointless. For instance, **Christoph**, the café owner talked about how planning is meaningless as total security does not exist:

> From start I do not have such great pension premises…I do not think so much about ageing, I do not know what is right, I did not organize anything … there is no 100% security for old age anyway, so I am not that motivated… (freelancer, insecure pension).

Lacking Resources from the State

Some precarious individuals expressed their concern that no matter how much one would try to prepare for old age as an individual, there are certain resources that need to come from the state. Some of these resources like cultural events where older people can participate for free or for a lower price. Several of the precarious and financially secure individuals mentioned they imagine they would participate in cultural life in their old age. Nevertheless, more than cultural events are needed to support a positive old age. **Thomas** talked about the need of a social system that supports older people who do not have financial means of survival:

> Each citizen should have some financial security, a minimum of financial security to keep him mobile, I mean active in a physical but also psychological sense and the material conditions to do something, start something up, be independent, not always on the limit of survival, that would change something within society (…) people need a legislative framework, a social framework to help them (…) when they are aging people need to keep active, they should be able to decide independently and for this people need a legislative and financial framework

Thomas and **Christoph** mentioned the importance of resources coming from the state, services that help with building social resources for the future.

Nevertheless, there are some state initiatives that help with active aging such as free cultural events for older people. **Kai** (freelance researcher in his early fifties) talked about the advantages of aging in a large city. In this respect, he mentioned the cultural events happening in the big city and hoped that these will stay the same when he will be older:

> nowadays many cultural events are financed or are at half price for older people. I hope it stays this way since in old age it is important to stay active, be stimulated, go to cultural events and keep social contacts. In (name of big city) there are many such programs and they are funded by the state, even public transport reductions for older people since these are also important for keeping them active. (…) a good health insurance would give me the feeling of security for old age, it is a psychological resource that keeps you going, makes you more relaxed for the future.

Planning Old Age Carefully: The Financially Secure Individuals

Financially secure individuals make a goal out of planning old age carefully and several participants mentioned the importance of starting as early as possible. For financially secure individuals, main resources for a positive old age include health, financial assets, and social engagement. Thus, participants talked about what they do to stay healthy in their old age, how they organize their financial assets for their retirement years, and how they plan to spend their time once they stop working. Their activities are organized like a "to do" list for reaching a positive old age. The preparation actions that financially secure individuals mention resemble the active aging policy discourse because they revolve around keeping a healthy body and mind and take responsibility for one's health and financial assets. In the following, I will discuss the strategies that were most often mentioned by financially secure individuals as good preparatory efforts for old age.

Plan Timely for a Good Old Age

Planning for old age represents a valuable resource for financially secure individuals. Planning can take the form of activities that one is doing at present and expect to continue until old age or new activities that one intends to start in the near future. Some financially secure participants talked about their hobbies that they want to continue after they retire or about new options they may consider once they are not working anymore. Retirement in this sense was regarded by most interviewees as an opportunity for doing things that one did not have time before. For example, some considered they have healthy routines and planned to continue with their healthy lifestyle. Others intended to take up new hobbies or engage in voluntary activities.

For example, **Florian**, a business consultant, talked about the hobbies he wants to pursue in the future. He believes that his current dancing hobby may not be appropriate when he is older and thus displays a negative view of aging:

> I took up dancing lessons again, but I know this not something I can do in my 80s, so I have to find something else, and also something that my next girlfriend will enjoy, and we can share this pastime. Dancing is something I might not be able to do when I am 70 or 80 but I may play golf. Or go sailing. Some friends of my family bought a boat and they are sailing 3 months every summer. Well maybe this is also not something to do when you are very old.

Discipline in terms of maintaining a healthy lifestyle or spending time wisely was also mentioned by financially secure individuals as a significant resource for a positive old age. For instance, **Alina**, a female pastor, in her fifties mentioned the relevance of self-discipline for positive aging:

> One needs a certain self-discipline to continue to engage in activities, to say it is hard but one will not give up.

In **Alina's**' view, discipline implies managing time and activities during the day. However, for her it also means respecting one's own program and wishes and keeping a certain structure (i.e., discipline) that one desires for oneself.

The right to have "time for oneself" came up in several interviews with financially secure individuals. For instance, **Sonia**, the owner of a small vintage shop and owner of real estate properties, talked about the freedom to enjoy and divide ones' time as one wants to. Even though she has her small shop where she sells vintage clothes and objects, her financial security is guaranteed by the real estate property she owns. Thus, she can afford to keep the shop for her own pleasure and spend her time as she pleases in her late life:

> You have to be a bit selfish with your time…if a friend is visiting me and I am tired for instance I would tell them I want to rest now…they understand, I want to conserve my energies for later…I do not need to do anything, only what is enjoyable

Train the Aging Body and Mind

Both physical and mental health were considered assets for a good old age. Training both body and mind were mentioned as strategies for reaching a good old age. For example, **Florian** the business consultant, started with a fitness program to improve his body condition and to be more fit in old age:

> Currently I try to avoid risk behaviour like drinking too much or have a fatty diet. I also started an intensive training program to improve my stamina, to have a healthy heart in old age. Muscle training, I did not start yet, but I plan to do that to prevent problems with my back when I get older

Several financially secure individuals mentioned they were engaging in physical activity regularly. This regular exercise gave them the feeling they are doing something for old age, and it also represented an activity that they enjoy. Nevertheless, apart from training the body, continuously developing the mind was mentioned as a strategy to reach a good old age.

For instance, **Carla**, a school teacher approaching retirement, is interested in keeping her mind fit and therefore started with mental training exercises that may prevent memory loss and improve concentration.

> I started to train my mind … some say you can't train your brain, but I believe you can, so I started to learn poetry by heart, I think it helps one keep the memory intact.

For others, both the physical and mental exercise are more unconventional. For example, **Jens**, a social worker in his late fifties, talked about his situation of being a father at this age. His children are very young, one is seven and one is four years old, so from his perspective, he could be their grandfather. **Jens** said that:

> for myself my children keep me fit, they are very lively and active, having children is a good anti-aging concept. A couple of years before I had some problems with my back but after my first child was born and I had to carry him and move a lot my back pain was gone, I really feel that I can do again all that I was doing when I was 30 years old, but of course it is not common to have such young kids at my age, but I enjoy it.

Training the body and mind is a strategy that lend support to existing literature on active aging perceptions (Bowling 2006; Jopp et al. 2014).

Secure Financial Independence

From the perspective of financially secure individuals, independence in old age can only be achieved when financial assets are secure. Some of the financially secure individuals said they do not trust the state pension system or state insurance. Thus, they already thought about a plan and implemented it concerning having financial or real estate assets for old age. For example, **Florian**, who is a successful business consultant said:

> Naturally I made investments since the pension from state is not a sure thing. I have some property and I can administrate my real estate businesses. This will also give me something to do in my old age, I will always have a task. It is important not to give away all the money that you make, but to put some aside, to have financial resources in old age. It is clear we are not safe with this pension system that cannot handle the aging society, so for me I have a mix of real estate, some assets in companies, my life insurance, so that if one does not function then I have real estate, or I have money from the life insurance

Fabian, who is the owner of real estate property, was even more pessimistic about the social security system concerning issues of aging. **Fabian** recommended avoiding the system altogether and building one's own resources.

> It is very bad to depend on the state, for something like your health insurance or care in old age, something like old people homes, these are ghettos for old people, heaven forbid! I am totally against something like this (…) the pension system leads to poverty in old age, I pity those people who still rely on it (…) me personally I never relied on these things, I would advise young people to just put money aside, invest in private insurance, spread your assets in different directions to have more than one financial resource.

Sonia, being someone who invested in real estate, can afford to live from the rent money she collects. Consequently, she has time for her little trivia shop. She came to Germany from her home country in the middle east and worked as a nurse initially. In Germany, she married and had three children, and therefore stayed home to care for her offspring. Together with her husband, they invested in assets like real estate. When she was getting older, she opened her own shop, not with the intentions to do business, but to be socially active and thus ensure a positive old age for herself:

> I opened the shop 27 years ago, it is really like my hobby, if it were my real job I would have to invest much more work, go to all the antiques markets to buy new things, sell…like this I can have the shop open 4 h every day, meet people, be socially engaged. I have some property and I can live from the rents and from this shop I make enough to pay for the space and combined with my other earnings I am ok. I have met very nice people through the shop, people that come in, they do not have to buy something, they come in, we chat, sometimes I go for coffee or dinner with some people I met through the shop. So, it is great, I am very grateful for my life

The vintage trivia shop keeps **Sonia** busy and involved. She always gets to meet new people and also has a regular clientele in the neighborhood that provides her with a sense of being socially connected.

Leave the Big City

An important theme that came up was the idea of moving out of the big city after retirement. Several financially secure individuals already have a country house, or they concretely make arrangements to move out of their city apartment. The move is explained either as looking to have more peace and quiet or just being close to nature. Another explanation concerns making "a fresh start" in old age.

For instance, **Clara** and her husband already have a house in the countryside where the husband who is retired goes regularly. He even spends some months a year at this country house. Once **Clara** also retires, they plan to sell their current house and move there whole time. Since the children are grownups, they do not need **Clara** and her husband to be there in the big city, which they associate with stress and work.

> We have this house in *region in Germany* and my husband who is retired already spends there several months per year and I go there during weekends. In winter he is with me in *big city* because he does not want to be alone there during winter months. when I retire, we will move there altogether, and I will engage in my own activities there. At the moment I do not have time, my days are filled with work at school in the morning and afternoons I need to regenerate, I really need to rest during the afternoon so that I have energy for the following day. (…) for my husband this house has been a dream for longer, I needed a bit of time to get used to life there, but now we have several social contracts there, all in all we have the house there for around thirty years and I can say I feel good there. Our house here in *big city* we will sell and leave the city.

Barbara and her husband have already made concrete plans about moving from the city within a couple of months after her retirement. The new location represents a sort of positive view of aging where they can concentrate on their interests and do something useful for others:

> We have made concrete arrangements to move to *region in Germany* where my son is living. There we can start fresh and we can focus on our intellectual interests, engage in physical activity, volunteer and above all stay healthy. (…) we leave Berlin, but we will stay in touch with our friends here and through our volunteer activities in *region in Germany* we will meet new people

Be Socially Active to Give Meaning to Old Age

Financially secure individuals mostly focus on giving back to the community and their generative role in old age. Social activism represents for them a way to provide old age with meaning. Financially secure individuals can afford to volunteer after they retire since they do not have to worry about financial problems. Some of the financially secure participants like **Barbara**, a school headmaster approaching retirement have already taken concrete steps for their volunteer project. **Barbara**'s husband has just retired, and they want to move to the countryside once she also retires. Their grownup children have their own family, but **Barbara** and her husband want to spend their time with community engagements:

> I have already started taking care of this in an active way, it implies social engagement. I do not want to spend my old age only with my family, I want to do something for the community, be engaged socially.

Barbara is proud about her professional achievements so far and would like to share this knowledge with the next generation.

In their search for positive meaning in old age, several financially secure participants mentioned social activities as the response to a positive old age. **Florian**, a successful business consultant, talked about social activism providing the opportunity to invest old age with meaning:

> I hope to have a mission in old age, to have either grandchildren to take care of or I will give talks to share my work knowledge with the next generations of business consultants (…) I believe mental health in old age is automatic when you have grandchildren to care for and then you have meaning in life, you always have something to do and then you are also mentally sane, you have a balanced life.

Similar to **Florian**, **Bruno** a successful business man, talked about the importance of having meaning in old age and how only social activity or engagement can provide this:

> For me financially is aging not a problem. The problem is how to find meaning for my time in old age. I can retire anytime, but the question is how will I fill my days? Should I do more sports or some charity work? Invest time in family and friends? For me the question is how will I use my time the right way?

All in all, in contrast to precarious individuals, financially secure persons have to worry more about "finding meaning" in old age as how to finance their old age (e.g., pay the rent, eat healthy, etc.). Because they can "afford" retirement, financially secure individuals can think about how they will spend the years when they do not have to work anymore. Overall, financially secure individuals feel more threatened by a meaningless old age as compared to precarious individuals who fear poverty and illness in old age.

References

Bowling, A. (2006). Lay perceptions of successful ageing: Findings from a national survey of middle aged and older adults in Britain. *European Journal of Ageing*, *3*(3), 123–136. https://doi.org/10.1007/s10433-006-0032-2.

Carraro, N., & Gaudreau, P. (2013). Spontaneous and experimentally induced action planning and coping planning for physical activity: A meta-analysis. *Psychology of Sport and Exercise, 14,* 228–248. https://doi.org/10.1016/j.psychsport.2012.10.004.

Carstensen, L. L., Fung, H. H., & Charles, S. T. (2003). *Motivation and Emotion, 27*(2), 103–123. https://doi.org/10.1023/a:1024569803230.

Foti, A. (2017). *General Theory of the Precariat. Great Recession, Revolution, Reaction*. Downloaded from http://networkcultures.org/publications.

Fung H. H. & Carstensen L. (2006). Goals change when life's fragility is primed: Lessons learned from Older Adults, the September 11 Attacks and Sars. *Social Cognition, 24*(3), 248–278.

Gebhardt, W. A. (2006). Contextualizing health behaviors: The role of personal goals. In De Ridder & de Witt (Eds.), *Self-regulation in health behavior* (pp. 25–43). John Wiley & Sons. https://doi.org/10.1002/9780470713150.ch2.

Gutheil, I. A., & Heyman, J. C. (2006). They don't want to hear us: Hispanic elders and adult children speak about end-of-life planning. *Journal of Social Work in End-of-Life & Palliative Care, 2*(1), 55–70.

Hilton, J. M., Gonzalez, C. A., Saleh, M., Maitoza, R., & Anngela-Cole, L. (2012). Perceptions of Successful Aging among Older Latinos, Cross-Cultural Context. *Journal of Cross-Cultural Gerontology, 27*(3), 183–199. https://doi.org/10.1007/s10823-012-9171-4.

Jopp, D. S., Wozniak, D., Damarin, A. K., De Feo, M., Jung, S., & Jeswani, S. (2014). How could lay perspectives on successful aging complement scientific theory? Findings from a U.S. and a german life-span sample. *The Gerontologist*, *55*(1), 91–106. https://doi.org/10.1093/geront/gnu059.

Lassen, A. J., & Moreira, T. (2014). Unmaking old age: Political and cognitive formats of active ageing. *Journal of Aging Studies, 30,* 33–46. https://doi.org/10.1016/j.jaging.2014.03.004.

Sheeran, P., Webb, T. L., & Gollwitzer, P. M. (2006). Implementation intentions: Strategic automatization of goal striving. In Denise T. D. de Ridder and John B. F. de Witt. (Eds.) *Self-Regulation in Health Behavior*, 119–145. John Wiley & Sons Ltd. https://doi.org/10.1002/9780470713150.ch6.

Standing, G. (2011). *The precariat: The new dangerous class*. London: Bloomsbury Academic.

Walker, A. (2009). Commentary: The emergence and application of active aging in Europe. *Journal of Aging & Social Policy, 21*(1), 75–93. https://doi.org/10.1080/08959420802529986.

Wolff, J. K., Warner, L. M., Ziegelmann, J. P., Wurm, S., & Kliegel, M. (2016). Translating good intentions into physical activity: Older adults with low prospective memory ability profit from planning. *Journal of Behavioral Medicine, 39*(3), 472–482. https://doi.org/10.1007/s10865-015-9707-5.

Part III

Chapter 8
Precarity and Views on Aging

Introduction

As discussed in the previous chapters, views on aging emerge in a social context (Wurm and Huxhold 2012; Craciun and Flick 2014; Fung 2013). Thus, precarity as a social setting, characterized by insecurity and uncertainty, may affect the views on aging of middle-aged persons who grow old in this social reality characterized by precarity. Following this line of thought, the present chapter aims to formulate some answers to the question "***how does precarity shape views on aging?***" Findings concerning the thoughts on aging of middle-aged people with precarious jobs and those who enjoy financial security will serve as the background for exploring the link between precarity and views on aging.

Several observations can be made concerning how precarity may shape views on aging. First, precarity fosters negative views on aging. These negative views on aging can be recognized in the various fears concerning aging that both precarious and financially secure individuals voiced during the interviews. Second, the interviewed precarious individuals depicted also positive experiences of aging. One can say that they adapted their positive views on aging to the social reality of precarity. In contrast, financially secure individuals have internalized the active aging policy discourses and their positive views on aging reflect these ideas. Third, the discrepancy between the active aging ideal and precarity as a social reality is not to be neglected. For precarious individuals, the contradictions between promoted ideals and social reality may induce the experience of a greater cognitive dissonance between ideal positive images of aging (i.e., discourses on aging) and their views on aging, constructed based on their daily life (i.e., precarious reality).

In the following I will analyze the negative views on aging fostered by precarity, as well as the positive views on aging that emerge despite living in a social reality characterized by precarity.

I. C. Crăciun, *Positive Aging and Precarity*, International Perspectives on Aging 21, https://doi.org/10.1007/978-3-030-14255-1_8

Negative Views on Aging and Precarity

As I have discussed in the first chapters of the book, the social reality of aging is influenced by policy and mass media discourses about positive aging (Lassen and Moreira 2014; Zimmermann and Grebe 2014; Stenner et al. 2010; Vickers 2007) as well as institutional responses to demographic changes (Carney and Gray 2015). In response to a global growth in aging population, the active aging policy recommends that people take responsibility for their own aging process and ensure well-being and health in old age (WHO 2002). While activity and social engagement are cherished, dependency and passivity become symbols of "unsuccessful agers" (Lassen and Moreira 2014; Minkler and Holstein 2008). Mass media messages are mixed since they present both images of active, happy, wealthy retirees (Gilleard and Higgs 2000) or even heroic old people (Carney and Gray 2015) and of sad, ill, and lonely elderly (Milner and Milner 2012; Rozanova 2010; Jolanki 2009). By promoting such contradictory images, mass media enhances both positive and negative images of old age. Aging individuals construct their identity based on such discordant accounts (Jolanki 2009) as well as their own experience with aging (their own or of significant others).

Even if people may think that they decide by themselves how they should live their old age, dominant social discourses on aging shape what it means to be an older person in a certain social context (Breheny and Stephens 2010). The experience of aging depends on social settings and material circumstances (Breheny and Stephens 2010). Precarity, by definition (Grenier et al. 2017; Portacolone 2013) inspires images of vulnerability, risk, dependency, and illness and thus may determine the development of negative views on aging. In precarious circumstances (e.g., precarious work, illness, etc.), negative views on aging might form the identity of aging individuals and thus make them engage less in preparations for an "uncertain" later life. Since neoliberal discourses (Katz and Marshall 2003) as well as active aging policies (Holstein and Minkler 2003) encourage individuals to take responsibility for arranging their old age, these may feel a great pressure to succeed. If they fail to reach the active aging goals (e.g., they do not engage in health promotion behavior, they do not make a pension plan), precarious workers will only have themselves to blame and this situation will enhance the precarity of their aging experiences. Thus, negative views on aging place precarious individuals in a self-fulfilling prophecy situation.

As discussed in previous chapters, within the German social context, the active aging discourse is present as theory (Baltes and Baltes 1990; Baltes and Carstensen 1996, 2003), as policy (Wurm and Huxhold 2012) and even in the form of lay people's views on successful aging (Jopp et al. 2015). However, findings from the interviews with precarious and financially secure middle-aged Germans showed that negative views on aging predominate.

According to the definition of negative views on aging, these imply a shortened perception of time left and no planning (Wurm et al. 2007; Wurm et al. 2010, 2013). Results from the present qualitative study point out that specifically for precarious individuals' negative images of aging comprise a short-time perspective and the feeling of "no future." The lack of planning may be connected to certain fears in addition to perceiving a limited time frame. Several fears concerning old age were associated with their experience of precarity, understood not only as financial precarity but also as a vulnerability of body and mind. Furthermore, the duality between body and mind was often mentioned as a characteristic of aging by the interviewees. Previous studies recognized such an artificial duality between body and mind (McHugh 2003; Andrews 1999) as an illustration of hidden negative stereotypes (e.g., not wanting to grow old at all but desiring to stay young).

Considering the characteristics of precarity (Grenier et al. 2017; Craciun and Flick 2014), it comes as no surprise that precarious individuals display negative views on aging rather than positive ones. Moreover, the situations when both precarious and financially secure individuals think about aging are directly linked to precarity (e.g., illness, vulnerability, insecurity). This suggests that settings characterized by precarity trigger negative automatic thoughts about aging such as "growing old means becoming ill and less attractive" or "old people are ill and dependent." The precarity of the body or the precarity of life in general (e.g., not finding a job, not having an apartment to live in, etc.) strengthen anxieties regarding aging.

Taking all the above into consideration, one can say that precarity represents a source for negative views on aging. Present data add to the existing literature by illustrating the link between precarity and aging, a link provided by views on aging. In the following, I will explore all the above-mentioned connections between precarity and negative views on aging in more depth. First, I will depict the fears that constitute an expression of precarity, then I will go on to discuss the precarious body as a symbol of aging and analyze the precarization of old age as a concept.

Fear of Old Age as an Expression of Precarity

Precarity manifests itself in the form of certain fears that both precarious and financially secure individuals have regarding old age. For example, findings from the present qualitative research showed that all participants feared disability, illness, dependency, and vulnerability. Since all of these are expressions of precarity, one may say, precarity finds itself at the core of participants' fears concerning old age. The very essence of precarity (i.e., vulnerability, insecurity, risk) is embedded in the fears associated with old age.

Precarity as a concept has been described in relation to certain locations such as work (Standing 2011), migration (Standing 2011, 2012; Kaida and Boyd 2011; Marier and Skinner 2008), living alone (Portacolone et al. 2018), or the aging as a process in itself (Grenier et al. 2017). In what concerns the experience of aging, financially secure individuals and precarious individuals share the same reality where social media discourses and active aging policy encourage people to thrive for a "positive old ag" (Rozanova 2010; McHugh 2003; Gilleard and Higgs 2008). Furthermore, they share the same context of precarity (e.g., the dissolution of the welfare state, the privatization of health services, changes in family relations, etc.). Thus, it is not surprising that financially secure and precarious workers share certain fears concerning old age, fears that make out their negative views on aging.

Upon examination of the interviews, one notices that views on aging were similar and mostly negative in both precarious and financially secure individuals. Instead of talking about hopes for the future, participants frequently voiced their fears concerning old age and aging. The fears concerning old age are quite similar since they emerge in the same social context of precarity, thus showing the importance of the latter. Precarity nurtures certain fears concerning old age such as financial dependency, vulnerability induced by worsening health, loneliness, loss of meaning of life. The fear of illness and disability is related to the representation of old age as an illness (see Chap. 7). The fear of becoming less attractive with age as well as being unproductive and therefore not useful is related to representing the aging body as vulnerable, frail or ugly (see Chap. 7).

As mentioned in Chap. 4, the meanings of precarity comprise the concept of uncertainty, of "being dependent of chance and circumstance," of "risk of falling, collapse, and similar accident" (Grenier et al. 2017; Portacolone et al. 2018). The definitions of precarity imply that it encompasses a feeling of uncertainty, insecurity, vulnerability, a perception of risk, as well as need and provision of care (Grenier et al. 2017). All these terms are often associated with aging, particularly with negative images of aging. According to the latter, old age is portrayed as a time of insecurity (e.g., financial) and vulnerability (e.g., due to illness, loss of body strength or loneliness). Therefore, one can say that the concept of precarity embodies to a certain extent negative images of aging. Despite existing positive discourses of aging (Mahne et al. 2017) and the dominance of successful aging theories (Liang and Luo 2012; Jopp et al. 2015), old age is usually conceptualized in terms of risk and vulnerability. This fact could be attributed to the role of the social context of precarity in shaping views on aging.

Not only aging as a process or old age as a stage in life, but also **the elderly** as a social group are directly related to precarity. "Old people" as a social group are often described in connection to precarity since they are presented as "a risk" to society or to themselves. The risk to society is embedded in what has been called the "apocalyptic demography" (Gee and Gutman 2000), while the individual risk to end up ill and alone in old age is regarded as a personal failure to adhere to the positive aging standard (Rozanova 2010). Findings from the present research showed that both precarious and financially secure individuals constructed "the elderly" as a precarious element in society. This lends support to earlier literature that portrays older individuals as a "threat" to society or a "silver tsunami" who endangers the equilibrium of social reality (Phillipson 2013; Barusch 2013). Such findings imply that society itself becomes precarious because of the increasing number of older people that "threaten" its balance and good functioning (Barusch 2013; Weicht 2013; Zimmermann and Grebe 2014). As evidence one can look at the disparity between the number of old people in need of care, and those who can offer care, or the deficiencies of the retirement systems based on pay-as-you go principles. For example, metaphors from German mass media and culture were shown to include suggestions about the burden of old age, "the mountain of retirees," or "dark clouds on the demographic horizon" (Grebe et al. 2013). The negative ideas behind such messages imply that there will be too many unproductive older adults and too few industrious individuals to sustain them financially and provide the needed care. These facts can only add to the precarity of existence for both older and younger individuals. Mass media discourses often emphasize the link between the progressively aging population and the danger of collapse of the health care system, pension system, or labor market (Weicht 2013). Furthermore, mass media highlights the personal responsibility toward achieving a "successful" old age (Rozanova 2010; Holstein and Minkler 2003; Minkler and Holstein 2008). People are left to fear the rising older population as well as the risk of becoming part of the "threat" at some point in their alter life. In this way, the awareness of the elderly as a threatening social group contributes to the maintenance of an awareness of precarity both at individual and social levels. All in all, fear itself constitutes a reflection of precarity, since it embodies uncertainty about something such as one's health or freedom to make choices. Since they grow up and grow old in a precarious reality dominated by negative aging stereotypes, people are socialized to fear old age, aging and the elderly.

The Precarious Body and Mind

Negative views on aging comprise the representations of the decaying older body and mind. With old age, the body losses its attractiveness and its functionality. With passing years, the body as well as the mind become less useful to the individual and to society. In this context precarity refers to the body or mind being constantly "at risk" (e.g., for illness, accidents), being vulnerable or in a state of collapse (e.g., one loses energy and vitality, one loses one's memory, etc.). The vulnerability of the human body represents both a negative view of aging and an illustration of precarity.

During the present research, the aging process was described in terms of experiencing symptoms of bodily and mental decline. Old age was constructed as a disease that manifests itself through physical and mental signs. Through the physical and psychological vulnerability that characterizes it, old age is represented as an expression of precarity. Because old age is associated with illnesses, it comes as no surprise that views on aging are embedded in ***fears*** concerning ***illness and dependency***. The description of old age almost as a disease that is made visible through physical signs points out the ***aging as disease metaphor***. Representing aging as a disease reflects a negative view on aging (Wurm et al. 2013). According to stereotype embodiment theory (Levy 2009), such negative views on aging are developed based on existing social representations of aging and personal experience with old age. Previous studies have pointed out that negative aging stereotypes prevail in Western cultures (Kotter-Grühn 2015; Wolff et al. 2014; Levy et al. 2002). For example, aging is linked with physical and social losses or perceiving oneself as becoming frail and dependent (Wolff et al. 2018). Aging is often associated with disease because it implies physiological changes and frequently a decline in health (Wurm et al. 2013). Unfortunately, this fact is not far from reality since older adults are more likely to suffer from a chronic disease (Yancik 2005) or to experience accidents, falls, and fractures (Fried 2000). Furthermore, old age was found to be associated with an increase in mental problems such as depression and anxiety (Freeman et al. 2016; Byers et al. 2010; Wolitzky-Taylor et al. 2010). Therefore, with an advanced age it becomes more difficult to distinguish illness from old age (Levy et al. 2009b). Nevertheless, the successful aging paradigm takes the view that aging is not automatically associated with disease and disability and distinguishes between pathological and successful aging. Negative self-perceptions of aging imply that people ascribe physical or mental losses to aging rather than to illness (Wurm et al. 2013; Sarkisian et al. 2002). Specific illness such as diabetes or arthritis are perceived as being caused by aging rather than other factors such as genetics or lifestyle (Wurm et al. 2013). However, attributing illness signs entirely to aging is considered dangerous, since it demotivates people from doing something to improve their condition. When faced with disability, people come to interpret it as being "normal" and accept it as a sign of old age. Previous research has found that older adults who attributed physical symptoms to aging showed little health behavior (Sarkisian et al. 2002, 2005; Gump et al. 2001) compared to adults who ascribed illness to other causes. According to study findings, if people saw aging as causing illness, they tended to perceive illnesses as less curable, preventable, and controllable (Sarkisian et al. 2001; Stewart et al. 2012).

Aging is intrinsically linked to physiological processes. These biological processes include a worsening of health but cannot be reduced to illness. However, there is a debate in the literature with some claiming that aging cannot be separated from disease (Gems 2015), while others argue for the existence of a "normal aging" where there is risk for illness, but the latter can be avoided (Hayflick 2004). Successful aging ideas can be integrated in the latter category. Such authors consider that attributing illness to aging is wrong. Thus, considering aging to be pathological represents a negative view on aging.

There is a body of evidence showing that self-perceptions of aging play a role in influencing health in old age (Wurm et al. 2007; Levy et al. 2002; Sargent-Cox et al. 2012) as well as longevity (Kotter-Grühn et al. 2009; Levy and Myers 2004, 2005). Self-perceptions of aging were shown to determine health changes rather than the other way around (Wurm et al. 2013; Sargent-Cox et al. 2012). This means that they influence health decline, they can become self-fulfilling prophecies. Among the psychological mechanisms explaining the relation between self-perceptions of aging and health, perceived control was identified (Levy et al. 2002). In case of precarious individuals, perceived control over their health and resources seems to be quite low and thus they are at a disadvantage in terms of ensuring their health as resource for old age. Even if they have positive views on aging, because they lack control over resources they would probably not invest in their health.

The association of aging with illness is more or less realistic since most people suffer from at least one chronical condition in their old age. The question remains whether we can call this representation a negative or rather a realistic view on aging that helps people prepare for old age. When knowing that illness will come, one tries to do something to prevent it. Thus, even if illness is not completely eliminated, at least the onset of disabilities is being pushed toward an "uncertain future," toward an older age. As pointed out by Grenier and her colleagues (2017), the "forth age" becomes the vulnerable age (and thus precarious), characterized by disability while the "third age" remains still a positive, active life stage. Nevertheless, the consideration of the third age as being "still young" and the forth age as a period of decline does not eliminate the problem, it just "postpones" it to a later stage in life. All things considered, old age should not be reduced to illness and more studies are needed to understand and clarify the relation between aging and disease.

As discussed above, precarity expresses the intrinsic vulnerability of the aging body and mind. Therefore, precarity does not affect views on aging just in terms of problems related to the insecurity of the labor market. When thinking about old age, middle-aged individuals do not automatically connect it with problems of work and retirement. Issues of cognitive and bodily impairment often take precedence on their list of aging-related worries. The first signs of vulnerability and disability related to aging are noticed in the body (e.g., people lose their mobility, start to feel muscle and joint pains, etc.) and sometimes in the mind (e.g., people start to forget things, they cannot learn new things as easily as before, etc.). Since aging is first recognized in the appearance of a persons' face and body, these become "the object" where precarity is embodied. According to the interviewed participants, in order to fight "aging," one needs to act on the body or on the mind. Mostly the body but every now and then also the mind needs to be altered in order to hide any signs of vulnerability and decline. The aging body and mind become the targets of prevention.

The advertisement industry was shown to construct aging as a disease and offers the "cure" in the form of anti-aging cosmetics. Smirnova (2012) calls these "cosmeceutical," because they combine cosmetics with pharmaceutical properties. The negative aging stereotypes that inspire such advertisements are associating old age with ugliness and disease, while "staying young" is linked with beauty and health. Feminist aging researchers challenge sayings such as "*fifty is the new thirty*" or "*fifty is still sexy*" because these expressions promote rigid and ageist standards of beauty, especially for older women. These standards are difficult to achieve in real life since they are mostly promoted and maintained by role models through repeated cosmetic surgeries. All signs of decline in an aging body (e.g., wrinkles, fat deposits, etc.) are made into arguments for selling anti-aging products to hopeful women and lately also men. These "symptoms" of aging encourage individuals to survey their bodies and identify aging. The body is perceived as an age identity card (Craciun and Flick 2014). Thus, it comes as no surprise that when asked to define old age, both precarious and financially secure individuals considered physical signs relevant and talked at length about their aging bodies. Looks and physical symptoms (e.g., pains, stiff muscles, and joints) are included on this "identity card" for old age. Psychological changes were also mentioned especially by precarious workers who seemed to be more afraid of losing their mental capacity to work and be productive in old age.

In relation to the aging body and mind, precarious middle-aged individuals focused mainly on their ***fears of being unproductive*** in old age, while financially secure persons focused on fears about ***diminishing health*** or ***loss of attractiveness***.

The neoliberal discourse implies that individuals have a civic duty to remain healthy, called by Rose "a will to health" (2001) and by Smirnova "a will to youth" (2012). According to the "successful aging" discourse, people must take responsibility for their health and well-being in old age. When aging individuals fail to stay healthy and young looking, they consider it mainly their own fault. The "self-responsibility" discourse is internalized so that precarious individuals mistrust institutions and rely more on themselves for old age than on external factors (e.g., other people, institutions, etc.). In a context of successful aging, the body becomes the first measure of success (Smirnova 2012) but also as present findings point out, a mirror for personal failure. For example, if a woman looks young, she is also considered healthy, while if she "looks older," she is considered ill (Holstein and Minkler 2003). When confronted with a reality characterized by precarity, individuals are reminded that they need to be responsible for their aging precarious body and mind. They must work to maintain their health and productivity in old age. The "successful aging" discourse that emphasizes personal responsibility enhances fears concerning old age and thus also indirectly strengthens negative views on aging.

Settings where people think about aging or situations they associate with "old age" are relevant for understanding the experiences people have with old age or older people. These experiences shape their personal views on aging. Thus, it is relevant to analyze settings, contexts, or situations that trigger thoughts concerning aging and old age. **Contexts** that caused thoughts about aging were quite similar for precarious and financially secure individuals. The interviewed middle-aged individuals monitored signs of aging in themselves, in their family members or within their circles of friends. Precarious workers often discussed aging as a problem that they shared with their friends. Several precarious individuals mentioned the pressure they felt at work to stay young and how comparison to younger work colleagues made them feel old. Middle age is known to be a turning point when people start to think about the balance between gains and losses (Baltes et al. 1999). Therefore, middle age itself constitutes a life stage that stimulates thoughts about aging and old age. The settings where both precarious and financially secure middle-aged individuals thought about growing old affected their self-perceptions of aging because they embodied direct experiences with aging. Since aging experiences implied mostly noticing signs of weakness in oneself or others, these also nurtured negative views on aging. Hence, in order to change views on aging, the settings that trigger such negative thought should also be altered.

All in all, by highlighting the vulnerability of the aging body and mind, by associating illness with old age and bringing forward negative contexts for thinking about old age (e.g., risk situations), precarity fosters negative views on aging.

The Precarization of Old Age

During the interviews, old age was repeatedly constructed as "insecure" or in terms of the present book, as precarious. Notions such as "precarious" of "precarity" do not refer just to uncertain living conditions for old age or to the vulnerability of the aging body and mind, but to the very essence of old age as a life stage. Both precarious and financially secure individuals contested old age in several ways. First, participants considered chronological age to be irrelevant as compared to a subjective age (e.g., you are only as old as you feel). Second, the interviewed middle-aged Germans denied the existence of old age as a notion in itself and constructed an "absence of old age" replacing it with the concept of agelessness (e.g., you are not old you are still young). Third, for precarious individuals the feeling of "no future" that they often experienced signifies the absence of a clear representation of old age (e.g., I do not think about old age). Respondents, especially precarious workers, both contested and feared the idea of aging, constructing "old age" as an "undefined" or "insecure" and thus "precarious" concept. The feeling of "no future" was also found in the literature to be a characteristic of precarity (Foti 2017).

The fact that chronological age is considered obsolete by both precarious and financially secure individuals represents an expression of the negative image of aging. This may imply that a subjective perception of one's own age is more important than chronological age (i.e., years lived). In the literature, this was called the "subjective age bias," meaning that "*the older you get the younger you feel*" (Weiss and Lang 2012). However, minimizing the importance of chronological age was interpreted as an attempt to manage the aging process. By controlling how one "feels about aging" and one's age, a person can transcend age (Andrews 1999). Even when one would be considered old according to the years of life lived, one still has the possibility to feel, act and look young or younger than one's chronological age. In case of precarious individuals, if they dissociate themselves from their own age group and feel younger, they may perceive a longer future time perspective. This can constitute a mechanism to protect themselves from the short-termism of precarity. Previous studies did not find this effect in middle-aged adults (Weiss and Lang 2012). However, the awareness of precarity may change this fact. For precarious middle-aged adults it can become relevant to dissociate themselves from older age groups and from the process of aging itself which they associate with negative stereotypes. Through this process of dissociation, they distance themselves from the negative aging stereotype and thus may regulate their negative emotions. Since precarity by definition implies lack of control, it is not surprising that precarious individuals need to find ways to control the environment. Denial and dissociation may serve as ways to gain control over their circumstances and experiences.

Previous studies have shown that as people grow old, they are susceptible to internalize negative aging stereotypes (Levy 2009; Wurm et al. 2013). Because old age is perceived to be negative (e.g., associated with being unattractive and senile), it represents a threat to a person's well-being. Thus, people feel the need to protect themselves from old age by dissociating themselves from it. They start to consider that others may grow old or are already old, but they "*still feel and act young*" (Weiss and Lang 2012). Age groups dissociation is presented as a self-protective strategy used by older individuals in order to prevent negative age stereotypes from becoming self-defining (Weiss and Lang 2012). For middle-aged individuals, negative age stereotypes may also represent a threat and they emphasize the ways in which one can transcend old age (i.e., real chronological age). Hiding or denying one's age represent ways in which people protect themselves from a negative age identity (Weiss and Lang 2012). People accept or avoid certain age identities depending on the cognitive representation of those identities (Ashmore et al. 2004; Arndt et al. 2002; Ellemers et al. 1997; Mussweiler et al. 2000). People want to belong to groups that offer a positive social identity, in this case they want to be associated with the youth not with seniors. Positive aging discourses have been criticized for attempting to eliminate old age as a life stage in itself (Andrews 1999; Katz and Marshall 2003; McDaniel 2005; Rozanova 2010). Furthermore, the idea that "*old age is obsolete*" or that it constitutes merely a psychological state have been described in the literature as representing a subtle form of ageism (Andrews 1999; McHugh 2003; Rozanova 2010).

As previous research has pointed out, one effect of the internalization of successful aging discourses is that people come to favor agelessness instead of growing old (Andrews 1999; Katz and Marshall 2003). This situation begs the question "does old age still exist or is it being replaced by an indefinite state of "still young"?" Some participants acknowledged that people from the current baby boomers aging generation "do not look their age." This implies that old age is associated with negative traits (e.g., disability, vulnerability), and people who" do not look" or "do not act their old age" are presented as positive examples of non-aging. Rozanova (2010) in her study points out that when people present the elderly as exceptions to the rule, they emphasize the negative images of aging because positive examples remain exceptions who are "still young" rather than being old. The successful aging rhetoric was often criticized for equating success with staying young (Smirnova 2012; Katz and Calasanti 2015; Liang and Luo, 2012). Participants from both groups tried to hide signs of aging and attempted to look and act younger than they actually were. The hidden thought is that "*being old is bad,*" adding evidence to the existing negative images of old age. Participants have internalized ideals of agelessness. The ideal of successful aging is not to age at all or at least to minimize the extent to which aging is made visible to others (Andrews 1999; Rozanova 2010).

Because aging is first made visible in the body (appearance and body functions), an artificial duality between body and mind emerges. Such a false duality between body and mind and its negative consequences for perceptions of aging has been discussed before in the literature (Andrews 1999). Andrews (1999) talks about the divide between people who "are old" and those who "only look old" and the split between the internal and external self when facing old age (i.e., the mask of aging theory). The duality between body and mind is expressed in the literature as "the mask of aging" (Andrews 1999). It implies that there is a tension between the external appearance and the internal feelings of an aging person. The latter subjective experience of aging is meant to take precedence as people age. Thus, how old one feels becomes more relevant than how old one looks (i.e., face, body). Those who

to be someone else, in this case someone younger, implies that people do not accept themselves as aging individuals.

All in all, the denial of aging appears to be characteristic for precarity. Negative views on aging are promoted in a context of precarity, and older adults are marginalized. Precarity is promoting ageism by supporting negative images of aging and by making positive views on aging seem unrealistic. Sometimes, even trying to talk about old age in a positive way results in a denial of aging. Subsequently, one needs to ask, "what does positive aging mean in a context characterized by precarity?" In the following, I will analyze how positive old age is being "reinvented" in a context of precarity.

Reinventing Positive Aging in a Precarious Context

Dealing with the dissonance between precarity and positive aging constitutes a relevant challenge for precarious individuals It is solved by reframing the negative views, adding positive components, changing goals/restricting the positive elements to fit the precarious reality.

Positive images of aging are adapted to "fit" the precarious context. For example, short-term thinking can be considered a strength in a precarious context, while planning could be a barrier for positive aging in precarious times. As discussed in Chap. 6, wisdom may be more important for precarious middle-aged individuals than just training the body to keep it fit for old age. "Typical" positive aging role models are replaced with precarious people who are aging in a good way and embody precarious positive aging models. Taking all these points into consideration, one can affirm that positive views on aging are reinvented in a precarious reality. This may happen because precarious individuals require novel and realistic positive images of aging in order to deal with the ***cognitive dissonance*** between the precarious reality and ideals promoted by the positive aging policy. In the following, I will discuss how positive views on aging are redesigned in a precarious context, by analyzing each of the points above.

Reframing the Negative Characteristics of Views on Aging

According to the theoretical definition, positive views on aging imply that aging is associated with further development and with carefully planning the future (Wurm et al. 2010). Perceiving aging related gains suggests that the future still holds positive meaning for aging individuals. The fact that aging persons continue to perceive an ample future time frame enables them to make plans for old age (Wurm et al. 2010; Wolff et al. 2014). However, according to this definition, only financially secure individuals would have positive views on aging because they perceive an ample time frame and organize their old age. A more generous future time frame is facilitated by the existence of a retirement plan or financial assets that protect people from financial worries. The structure provided by retirement and financial resources enable financially secure individuals to set long-term goals for old age and plan how to achieve them. The goals that financially secure persons set for old age reflect the active aging policy philosophy, namely the wish to be independent in old age, healthy, and socially engaged (WHO 2002; Foster and Walker 2014).

Findings showed that when people reach the age of 50 years old, they experience a subjective threshold where they become aware that their future may be shorter than their past. This realization in itself is already responsible for the perception of a shorter time perspective. On top of this, precarity means perceiving a limited future or the feeling of "no future" as I have discussed in Chap. 4. Previous literature attests the feeling of "no future" as a characteristic of precarity (Standing 2011; Foti 2017). Precarity contributes to the awareness of a limited time frame as well as to a focalization on the present in precarious individuals. While a short-time perspective may constitute a component of negative views on aging in financially secure individuals, it might take on different meanings for precarious individuals. On the one hand, a shortened future time perspective may signify the awareness of the future uncertainty and be a component of a negative view of old age. On the other hand, it may point out that precarious individuals have different positive views of aging as compared to financially secure individuals. If long-term planning does not make sense in a precarious reality, focusing on short-term plans and building on resources can represent better strategies for active aging. Thus, a short-term time perspective can become part of a positive view on aging for precarious individuals. Nevertheless, even if the short-time perspective can be reframed as a component of positive views on aging, it does not mean that it will not affect health behavior practices, just as depicted in previous studies (Wurm et al. 2010). It is known that a short-time perspective and a predominant focus on the present are usually associated with less engagement in health behavior (Hall and Fong 2003). Eventually, such a short-time perspective may have detrimental effects on health and longevity (Levy et al. 2002; Levy 2009; Kotter-Grühn et al. 2009). All things considered, short-term thinking that originally represents a component of negative views on aging is reinterpreted as a positive view on aging in precarious times.

Adding New Components to Positive Aging Views

Precarious individuals also desire a good old age and have certain views that suggest one can experience life satisfaction in late life. However, new components may be added to positive views on aging in precarious contexts (e.g., when people feel vulnerable or at risk), while other common ingredients of positive aging (e.g., activity, planning, etc.) may be redundant (as discussed above). For example, activity for precarious individuals comes as "a given" and does not need to be set as a specific goal. Since work is required in order to earn a living in old age, passivity, or inactivity might prove fatal for precarious middle-aged people. Precarious workers focus on gathering resources for old age in the present and "store" them for potential future use. Living in the moment, making the most of present assets to learn skills for the future represents a favored strategy used by precarious workers.

Despite not planning, precarious workers have the expectation that their old age is associates with further development. Wisdom was considered a relevant component of positive aging and a reflection of personal development. This implies that precarious workers do foresee continuous development for the future that may eventually turn into wisdom. Precarious participants often mentioned that they can better solve problems now that they are older, and this represents for them a "sign of acquired wisdom." In this sense, most interviewed precarious participants saw aging itself as an accomplishment. In contrast, most financially secure individuals worried about lack of meaning related to old age and retirement. For precarious people, the future is just a continuation of the present where they hope their resources will last long enough to keep them healthy and independent. There is no clear threshold such as entering retirement to influence changes in their lives.

The positive views of aging or precarious individuals fit better within the SOC model (Baltes and Baltes 1990). Precarious people know they need to compensate for lack of financial resources and they work to optimize their skills at present and build their resources for the future. The resources that precarious individuals develop are mostly of a social nature. For instance, precarious middle-aged Germans see themselves surrounded by family or friends in old age. However, developing their social networks is not just something they plan for the future but what they actively presently pursue. They cultivate their family relations and their friendships. Work is another domain where they are careful to nurture their networks and enhance their job-related skills. They are aware that their future employment opportunities will depend on the social networks they build at present.

Finding novel components for positive aging representations concerns identifying positive models for aging among precarious workers. The examples presented by precarious and financially secure individuals in order to describe positive aging role models were portrayed as "exceptions" to the rule. Old people were described rather negatively, in accordance with the predominant negative aging stereotypes found in the Western world (Levy 2009). In contrast, examples of positive aging "break the cliché" and are "still active," "still healthy," and "still independent." For financially secure individuals, positive aging ideals center around being active, sporty, and socially engaged (e.g., volunteering for the community). For example, as some participants mentioned they do not want to spend time just being around the family or engaged in passive hobbies like fishing. Both financially secure and precarious individuals realized that they will not be "the granny with rosy cheeks" who go on bus tour holidays. Precarity stimulates a reinvention of ideal positive aging models. Precarious people wish for relaxation and time to enjoy life in old age. Wisdom and tranquility are on their wish list rather than sport activities or volunteering. All in all, wisdom and management of resources become part of the novel positive aging view for precarious individuals and novel role models are searched for in order to confirms the newly created positive representations.

The desire to make sense of life and maintain meaning over the lifespan represents a basic human need (Stevens et al. 2006). From childhood on, individuals use social artifacts to make sense of life, such as school, work, gender roles, retirement (Kegan 1994). In order to resolve conflicts that arise between societal demands and personal desires, they internalize social demands and transform them into personal goals (Fung 2013). The same may be true when people need to solve conflicts between social demands and the limited opportunities of their realities. Thus, so as to solve the conflict between positive aging recommendations and the daily hassles of precarity, middle-aged individuals internalize the positive aging goals but change them into their own version of aging views.

Individuals grow old in a particular culture and they make sense of what is meaningful in life by internalizing the values from that culture (Fung 2013). In the western world, success and independence are important values. This applies to aging representations as well. Theoretical models and the active aging policy encourage people to be successful in old age. Successful aging individuals should stay active, independent, and socially engaged. Nevertheless, for some persons, these goals contradict their daily reality characterized by uncertainty and insecurity. For example, precarious people face many barriers when they set goals to live independent lives in their old age. The discrepancy between what they are encouraged to wish for (i.e., positive aging goals) and what seems realistic in the given circumstances is greater than for financially secure individuals. The latter possess the financial resources to make an active old age happen. In order to deal with the discrepancy between goals and reality, precarious individuals have several options. They either reformulate their positive views on aging (i.e., formulate new goals for old age), refine their preparation efforts (see Chap. 9), or they just settle for a negative image of aging. All of these constitute different pathways of dealing with the stress induced by having to adopt a positive view on aging that does not resemble the reality of precarious aging. These strategies are ways of dealing with the experience of precarity. In the present chapter, I have focused on views on aging, while in Chap. 9, I will go on to talk about the preparations for old age in precarious conditions.

References

Andrews, M. (1999). The seductiveness of agelessness. *Ageing & Society, 19*(3), 301–318. https://doi.org/10.1017/s0144686x99007369.

Arndt, J., Greenberg, J., Schimel, J., Pyszczynski, T., & Solomon, S. (2002). To belong or not to belong, that is the question: Terror management and identification with gender and ethnicity. *Journal of Personality and Social Psychology, 83,* 26–43.

Ashmore, R. D., Deaux, K., & McLaughlin-Volpe, T. (2004). An organizing framework for collective identity: Articulation and significance of multidimensionality. *Psychological Bulletin, 136,* 80–114.

Baltes, P. B., & Baltes, M. M. (Eds.). (1990). *Successful aging*. Cambridge University Press. https://doi.org/10.1017/cbo9780511665684.

Baltes, M. M., & Carstensen, L. L. (1996). The process of successful ageing. *Ageing and Society, 16*(4), 397–422. https://doi.org/10.1017/s0144686x00003603.

Baltes, M. M., & Carstensen, L. L. (2003). The process of successful aging: Selection, optimization, and compensation. *Understanding Human Development*, 81–104. https://doi.org/10.1007/978-1-4615-0357-6_5.

Baltes, P. B., Staudinger, U. M., & Lindenberger, U. (1999). LIFESPAN PSYCHOLOGY: Theory and application to intellectual functioning. *Annual Review of Psychology, 50*(1), 471–507. https://doi.org/10.1146/annurev.psych.50.1.471.

Barusch, A. S. (2013). The aging tsunami: Time for a new metaphor? *Journal of Gerontological Social Work, 56*(3), 181–184. https://doi.org/10.1080/01634372.2013.787348.

Breheny, M., & Stephens, C. (2010). Ageing in a material world. *New Zealand Journal of Psychology, 39*(2), 41–48.

Byers, A. L., Yaffe, K., Covinsky, K. E., Friedman, M. B., & Bruce, M. L. (2010). High occurrence of mood and anxiety disorders among older adults: The national comorbidity survey replication. *Archives of General Psychiatry, 67*(5), 489–496.

Carney, G. M., & Gray, M. (2015). Unmasking the "elderly mystique": Why it is time to make the personal political in ageing research. *Journal of Aging Studies, 35,* 123–134. https://doi.org/10.1016/j.jaging.2015.08.007.

Craciun, C., & Flick, U. (2014). "I will never be the granny with rosy cheeks": Perceptions of aging in precarious and financially secure middle-aged Germans. *Journal of Aging Studies, 29,* 78–87. https://doi.org/10.1016/j.jaging.2014.01.003.

Ellemers, N., Spears, R., & Doosje, B. (1997). Sticking together or falling apart: ingroup identification as a psychological determinant of group commitment versus individual mobility. *Journal of Personality and Social Psychology, 72,* 617–626.

Foster, L., & Walker, A. (2014). Active and Successful Aging: A European Policy Perspective. *The Gerontologist, 55*(1), 83–90. https://doi.org/10.1093/geront/gnu028.

Foti, A. (2017). *General Theory of the Precariat. Great Recession, Revolution, Reaction.* Retrieved from http://networkcultures.org/publications.

Fried, L. P. (2000). Epidemiology of aging. *Epidemiologic Reviews, 22,* 95–106. https://doi.org/10.1093/oxfordjournals.epirev.a018031.

Freeman, A. T., Santini, Z. I., Tyrovolas, S., Rummel-Kluge, C., Haro, J. M., & Koyanagi, A. (2016). Negative perceptions of ageing predict the onset and persistence of depression and anxiety: Findings from a prospective analysis of the Irish Longitudinal Study on Ageing (TILDA). *Journal of Affective Disorders, 199,* 132–138. https://doi.org/10.1016/j.jad.2016.03.042.

Fung, H. H. (2013). Aging in Culture. *The Gerontologist, 53*(3), 369–377. https://doi.org/10.1093/geront/gnt024.

Gee, E. M., & Gutman, G. M. (2000). *The overselling of population aging: Apocalyptic demography, intergenerational challenges, and social policy*. New York: Oxford University Press.

Gems, D. (2015). The aging-disease false dichotomy: understanding senescence as pathology. *Frontiers in Genetics*, 6. https://doi.org/10.3389/fgene.2015.00212.

Gilleard, C., & Higgs, P. (2000). *Cultures of ageing: Self, citizen and the body*. London: Prentice Hall.

Gilleard, C., & Higgs, P. (2008). Promoting mental health in later life. *Aging and Mental Health, 12*(3), 283–284. https://doi.org/10.1080/13607860802102407.

Grebe, H., Otto, W.-G., & 7 Zimmermann, H.-P. (2013). "The journey into the land of forgetfulness". Metaphors of old age and dementia in the media. In U. Kriebernegg, R. Maierhofer, and H. Mörtl (Eds.), The ages of live. Living and aging in conflict?. *Aging Studies*(Vol. 3, pp. 89–106). Bielefeld: Transcript.

Grenier, A., Phillipson, C., Laliberte Rudman, D., Hatzifilalithis, S., Kobayashi, K., & Marier, P. (2017). Precarity in late life: Understanding new forms of risk and insecurity. *Journal of Aging Studies, 43,* 9–14. https://doi.org/10.1016/j.jaging.2017.08.002.

Gump, B. B., Matthews, K. A., Scheier, M., Schulz, R., Bridges, M. W., & Magovern, G. J. (2001). Illness representations according to age and effects on health behaviors following coronary artery bypass graft surgery. *Journal of the American Geriatrics Society, 49,* 284–289. https://doi.org/10.1046/j.1532-5415.2001.4930284.x.

Hall, P. A., & Fong, G. T. (2003). The effects of a brief time perspective intervention for increasing physical activity among young adults. *Psychology and Health, 18*(6), 685–706. https://doi.org/10.1080/0887044031000011044.

Hayflick, L. (2004). Debates: The Not-So-Close Relationship Between Biological Aging and Age-Associated Pathologies in Humans. *The Journals of Gerontology Series A: Biological Sciences and Medical Sciences, 59*(6), B547–B550. https://doi.org/10.1093/gerona/59.6.b547.

Holstein, M. B., & Minkler, M. (2003). Self, society, and the "new gerontology". *The Gerontologist, 43*(6), 787–796.

Jolanki, O. H. (2009). Agency in talk about old age and health. *Journal of Aging Studies, 23*(4), 215–226. https://doi.org/10.1016/j.jaging.2007.12.020.

Jopp, D. S., Wozniak, D., Damarin, A. K., De Feo, M., Jung, S., & Jeswani, S. (2015). How could lay perspectives on successful aging complement scientific theory? Findings from a U.S. and a German life-span sample. *The Gerontologist, 55*(1), 91–106. https://doi.org/10.1093/geront/gnu059.

Kaida, L., & Boyd, M. (2011). Poverty variations among the elderly: The roles of income security policies and family co-residence. *Canadian Journal on Aging / La Revue Canadienne Du Vieillissement, 30*(01), 83–100. https://doi.org/10.1017/s0714980810000814.

Katz, S., & Marshall, B. (2003). New sex for old: Lifestyle, consumerism, and the ethics of aging well. *Journal of Aging Studies, 17*(1), 3–16.

Katz, S., & Calasanti, T. (2015). Critical perspectives on successful aging: Does it "Appeal More Than It Illuminates"? *The Gerontologist, 55*(1), 26–33. https://doi.org/10.1093/geront/gnu027.

Kegan, R. (1994). *In over our heads*. Cambridge, MA: Harvard University Press.

Kotter-Grühn, D. (2015). Changing negative views of aging: Implications for intervention and translational research. *Annual Review of Gerontology and Geriatrics, 35*(1), 167–186. https://doi.org/10.1891/0198-8794.35.16.

Kotter-Grühn, D., Kleinspehn-Ammerlahn, A., Gerstorf, D., & Smith, J. (2009). Self-perceptions of aging predict mortality and change with approaching death: 16-year longitudinal results from the Berlin Aging Study. *Psychology and Aging, 24,* 654–667. https://doi.org/10.1037/a0016510.

Lassen, A. J., & Moreira, T. (2014). Unmaking old age: Political and cognitive formats of active ageing. *Journal of Aging Studies, 30,* 33–46. https://doi.org/10.1016/j.jaging.2014.03.004.

Levy, B. R., Slade, M. D., & Kasl, S. V. (2002). Longitudinal benefit of positive self-perceptions of aging on functional health. In: *Journal of Gerontology*, *57*(5), 409–417.

Levy, B. R., & Myers, L. M. (2004). Preventive health behaviours influenced by self-perceptions of aging. *Preventive Medicine, 39,* 625–629.

Levy, B. R., & Myers, L. M. (2005). Relationship between respiratory mortality and self-perceptions of aging. *Psychology and Health, 20,* 553–564. https://doi.org/10.1080/14768320500066381.

Levy, B. R., Ashman, O., & Slade, M. D. (2009a). Age attributions and aging health: Contrast between the United States and Japan. *The Journals of Gerontology Series B: Psychological Sciences and Social Sciences, 64B*(3), 335–338. https://doi.org/10.1093/geronb/gbp002.

Levy, B.R., Zonderman, A., Slade, M.D., & Ferrucci, L. (2009b). Negative age stereotypes held earlier in life predict cardiovascular events in later life. *Psychological Science, 20,* 296–298.

Levy, B. (2009). Stereotype embodiment. *Current Directions in Psychological Science, 18*(6), 332–336. https://doi.org/10.1111/j.1467-8721.2009.01662.x.

Liang, J., & Luo, B. (2012). Toward a discourse shift in social gerontology: From successful aging to harmonious aging. *Journal of Aging Studies, 26*(3), 327–334. https://doi.org/10.1016/j.jaging.2012.03.001.

Mahne, K., Wolff, J. K., Simonson, J., & Tesch-Römer, C. (2017). Altern im Wandel: Zwei Jahrzehnte Deutscher Alterssurvey [Changes in Aging: Two Decades German Survey of Aging]. In K. Mahne et al. (Eds.) *Altern Im Wandel*, 11–28. https://doi.org/10.1007/978-3-658-12502-8_1.

Marier, P., & Skinner, S. (2008). The impact of gender and immigration on pension outcomes in Canada. *Canadian Public Policy, 34*(4), S59–S78. https://doi.org/10.3138/cpp.34.4.s59.

McDaniel, S. A. (2005). Sentenced by a metaphor: Living with aging. *Recent Advances, Research Updates, 6*(3), 275–278.

McHugh, K. E. (2003). Three faces of ageism: Society, image and place. *Ageing & Society, 23*(02), 165–185. https://doi.org/10.1017/s0144686x02001113.

Milner, C., Van Norman K., & Milner, J. (2012). Media's portrayal of ageing. In *Global Population Ageing: Peril or Promise?* Retrieved from http://demographic-challenge.com/files/downloads/6c59e8722eec82f7ffa0f1158d0f4e59/ageingbook_010612.pdf.

Minkler, M., & Holstein, M. B. (2008). From civil rights to … civic engagement? Concerns of two older critical gerontologists about a "new social movement" and what it portends. *Journal of Aging Studies, 22*(2), 196–204. https://doi.org/10.1016/j.jaging.2007.12.003.

Mussweiler, T., Gabriel, S., & Bodenhausen, G. V. (2000). Shifting social identities as a strategy for deflecting threatening social comparisons. *Journal of Personality and Social Psychology, 79,* 398–409.

Phillipson, C. (2013). *Ageing* (218 pp.). Cambridge: Polity Press, ISBN 13: 978 0 7456 3085 4.

Portacolone, E. (2013). The notion of precariousness among older adults living alone in the U.S. *Journal of Aging Studies, 27*(2), 166–174. https://doi.org/10.1016/j.jaging.2013.01.001.

Portacolone, E., Rubinstein, R. L., Covinsky, K. E., Halpern, J., & Johnson, J. K. (2018). The Precarity of older adults living alone with cognitive impairment. *The Gerontologist*. https://doi.org/10.1093/geront/gnx193.

Rozanova, J. (2010). Discourse of successful aging in The Globe and Mail: Insights from critical gerontology. *Journal of Aging Studies, 24*(4), 213–222. https://doi.org/10.1016/j.jaging.2010.05.001.

Sargent-Cox, K. A., Anstey, K. J., & Luszcz, M. A. (2012). The relationship between change in self-perceptions of aging and physical functioning in older adults. *Psychology and Aging, 27,* 750–760. https://doi.org/10.1037/a0027578.

Sarkisian, C. A., Liu, H., Ensrud, K. E., Stone, K. L., Mangione, C. M. For The Study, o., & Group, O. F. R. (2001). Correlates of attributing new disability to old age. *Journal of the American Geriatrics Society, 49,* 134–141. https://doi.org/10.1046/j.1532-5415.2001.49033.x.

Sarkisian, C. A., Hays, R. D., & Mangione, C. M. (2002). Do older adults expect to age successfully? The association between expectations regarding aging and beliefs regarding healthcare seeking among older adults. *Journal of the American Geriatrics Society, 50,* 1837–1843. https://doi.org/10.1046/j.1532-5415.2002.50513.x.

Sarkisian, C. A., Prohaska, T. R., Wong, M. D., Hirsch, S., & Mangione, C. M. (2005). The relationship between expectations for aging and physical activity among older adults. *Journal of General Internal Medicine, 20,* 911–915. https://doi.org/10.1111/j.1525-1497.2005.0204.x.

Smirnova, M. H. (2012). A will to youth: The woman's anti-aging elixir. *Social Science and Medicine, 75*(7), 1236–1243. https://doi.org/10.1016/j.socscimed.2012.02.061.

Standing, G. (2011). *The Precariat: The new dangerous class*. London: Bloomsbury Academic.

Standing, G. (2012). The precariat: From denizens to citizens? *Polity, 44*(4), 588–608.

Stevens, N. L., Martina, C. M. S., & Westerhof, G. J. (2006). Meeting the need to belong: Predicting effects of a friendship enrichment program for older women. *The Gerontologist, 46*(4), 495–502. https://doi.org/10.1093/geront/46.4.495.

Stewart, T. L., Chipperfield, J. G., Perry, R. P., & Weiner, B. (2012). Attributing illness to "old age": Consequences of a self-directed stereotype for health and mortality. *Psychology and Health, 27,* 881–897. https://doi.org/10.1080/08870446.2011.630735.

Stenner, P., McFarquhar, T., & Bowling, A. (2010). Older people and "active ageing": Subjective aspects of ageing actively. *Journal of Health Psychology, 16*(3), 467–477. https://doi.org/10.1177/1359105310384298.

Vickers, K. (2007). Aging and the media: Yesterday, today, and tomorrow. *Californian Journal of Health Promotion, 5*(3), 100–105.

Weicht, B. (2013). The making of "the elderly": Constructing the subject of care. *Journal of Aging Studies, 27*(2), 188–197. https://doi.org/10.1016/j.jaging.2013.03.001.

Weiss, D., & Lang, F. R. (2012). "They" are old but "I" feel younger: Age-group dissociation as a self-protective strategy in old age. *Psychology and Aging, 27*(1), 153–163. https://doi.org/10.1037/a0024887.

WHO. (2002). *Active ageing: A policy framework*. Geneva: Author.

Wolff, J. K., Warner, L. M., Ziegelmann, J. P., & Wurm, S. (2014). What do targeting positive views on ageing add to a physical activity intervention in older adults? Results from a randomised controlled trial. *Psychology and Health, 29,* 915–932. https://doi.org/10.1080/08870446.2014.896464.

Wolff, J. K., Beyer, A.-K., Wurm, S., Nowossadeck, S., & Wiest, M. (2018). Regional impact of population aging on changes in individual self-perceptions of aging: Findings from the German ageing survey. *The Gerontologist, 58,* 47–56. https://doi.org/10.1093/geront/gnx127.

Wolitzky-Taylor, K. B., Castriotta, N., Lenze, E. J., Stanley, M. A., & Craske, M. G. (2010). Anxiety disorders in older adults: A comprehensive review. *Depression and Anxiety, 27*(2), 190–211. https://doi.org/10.1002/da.20653.

Wurm, S., & Huxhold, O. (2012). Sozialer Wandel und individuelle Entwicklung von Altersbildern [Social change and individual devolvement of views on aging]. In F. Berner et al. (Eds.), *Individuelle und Kulturelle Altersbilder* [Individual and cultural views on aging], 27–69. https://doi.org/10.1007/978-3-531-93286-6_2.

Wurm, S., Tesch-Römer, C., & Tomasik, M. J. (2007). Longitudinal findings on aging-related cognitions, control beliefs, and health in later life. *The Journals of Gerontology, Series B: Psychological Sciences and Social Sciences, 62,* 156–164. https://doi.org/10.1093/geronb/62.3.P156.

Wurm, S., Tomasik, M. J., & Tesch-Römer, C. (2010). On the importance of a positive view on ageing for physical exercise among middle-aged and older adults: Cross-sectional and longitudinal findings. *Psychology and Health, 25*(1), 25–42. https://doi.org/10.1080/08870440802311314.

Wurm, S., Warner, L. M., Ziegelmann, J. P., Wolff, J. K., & Schüz, B. (2013). How do negative self-perceptions of aging become a self-fulfilling prophecy? *Psychology and Aging, 28*(4), 1088–1097. https://doi.org/10.1037/a0032845.

Yancik, R. (2005). Population aging and cancer: A cross-national concern. *The Cancer Journal, 11,* 437–441. https://doi.org/10.1097/00130404-200511000-00002.

Zimmermann, H.-P., & Grebe, H. (2014). "Senior coolness": Living well as an attitude in later life. *Journal of Aging Studies, 28,* 22–34. https://doi.org/10.1016/j.jaging.2013.11.002.

Chapter 9
Precarity and Preparing for Old Age

Introduction

Views on aging are known to influence preparation for old age (Wolff et al. 2014, 2016; Wurm et al. 2010, 2013; Levy 2009). Furthermore, views on aging emerge in a social context (Fung 2013; Breheny and Stephens 2010) and preparations for aging also happen within a social reality. In chapter eight, I analyzed how precarity as a social context of aging shapes views that individuals hold concerning old age. In the present chapter, I will continue this line of argument and attempt to answer the question "***how does precarity shape preparations for old age***?"

The comparison between precarious and financially secure individuals helps distinguish between different preparatory actions and thus explore the role of the social context of precarity for individual preparatory actions. In order to understand how precarity may affect preparations for old age, one needs to study the link between active aging policy (Lassen and Moreira 2014; Foster and Walker 2014) and the notion of precarity (Grenier et al. 2017; Portacolone et al. 2018). Active aging policy not only delineates what successful aging means, but it also makes recommendations concerning "good preparations" for aging well. Therefore, I first explore how the active aging discourse intertwines with precarity, making the point that precarity shapes the way people plan or fail to plan their old age. Then, I move on to analyze how precarity affects the context of aging preparations, mental preparations for old age, as well as shaping new resources for aging well.

According to the active aging policy framework, positive aging means providing the opportunities for health, participation, and security that would ensure quality of life as people age (WHO 2002; Stenner et al. 2010). However, some of these pillars of the active aging framework are difficult to apply in a context of precarity. Findings from the present qualitative study point out that opportunities for health are rare for precarious individuals, security represents the opposite of precarity, and participation or engagement are difficult in precarious times.

Both precarity (Grenier et al. 2017) and active aging (Foster and Walker 2014; Walker 2009; WHO 2002) represent complex concepts that influence the social reality

I. C. Crăciun, *Positive Aging and Precarity*, International Perspectives on Aging 21, https://doi.org/10.1007/978-3-030-14255-1_9

where people grow old. Precarity, as I discussed also in previous chapters, refers to a feeling of uncertainty and insecurity (Grenier et al. 2017; Portacolone et al. 2018). It has been used in the literature to illustrate an increasing lack of security in the context of global social and economic change (Grenier et al. 2017), insecurities on the labor market (Standing 2011), growing inequalities (Biggs 2014) or aging, and exclusion (Craciun and Flick 2014, 2015). The active aging policy characterizes a multifaceted reality (Lassen and Moreira 2014; Moulaert and Paris 2013) since there are different ways to reach a good old age. In similar ways, precarity represents a complex reality, a many-sided phenomenon that still needs to be defined and better understood.

The framework of active aging and the reality of precarity share the neoliberal ideology of success (Liang and Luo 2012; Katz and Marshall 2003), however, in different ways. Success represents an outcome for active aging policies (Foster and Walker 2014), while for precarity it constitutes a necessity for survival. Whoever is not successful in precarious times is at risk of physically disappearing or of being a victim of social exclusion. Since precarity embodies new forms of risk for aging individuals, these need to overcome all the encountered challenges and survive in precarious times. The thrive to be productive and engaged characterizes successful aging theories and the active aging ideology (Liang and Luo 2012; Minkler and Holstein 2008) but also precarity. Passivity is not an acceptable trait since it has negative consequences. Inactivity may lead to either an unhappy old age or failure to survive in precarious conditions. Individual responsibility is cherished both by active aging principles (e.g., the individual is responsible for his or her own lifestyle that may lead to healthy aging) and by precarious work conditions (e.g., the individual is responsible for finding a job, for retirement benefits). All in all, the active aging policy requirements are in conflict with the conditions of the precarious reality. For example, the capacity for agency, an important component of active aging guidelines, is limited by the conditions of precarity. Even if precarious people want to be socially engaged and productive, when work and health are uncertain, their good intentions may not be transformed into practice. In this sense, precarity creates an obstacle for achieving a good old age.

How people prepare for old age is shaped by policy and the social context where individuals grow old (Breheny and Stephens 2010). Positive aging policy recommends that people change their lifestyles to age in a healthy way (WHO 2002) and that both institutions and expectations are altered to promote a good old age (Lassen and Moreira 2014). Starting from the main pillars of the active aging framework (WHO 2002; Stenner et al. 2010), a good preparation for old age would include creating opportunities for health, benefiting from security and participation in old age.

On the whole, active aging policy ideas are not that different from what the interviewed middle-aged individuals desire. Especially financially secure individuals have internalized the ideals of the active aging policy and prepare for old age accordingly. Precarious middle-aged interviewed participants acknowledged the importance of preparing for an active old age. However, they talked a lot about the barriers they perceived when attempting to prepare for old age. Nevertheless, they also created their own resources that may help them prepare for old age in precarious conditions. In the following, I will explore how precarity shapes preparations for old age by referring to how it outlines the context where preparatory efforts are performed, how it affects mental preparations and how it fosters the development of novel resources for old age. First, I will analyze how precarity shapes an inadequate environment for preparations since it encourages the accumulation of health inequalities, nurtures ageism in work environments, and fosters mental barriers concerning preparation. Second, I will discuss how alternative preparation efforts are invented for old age.

Precarity and the Accumulation of Health Inequalities

According to active aging policies, individuals are encouraged to make healthy choices throughout their life span, remain active, and avoid illness and dependency (WHO 2002; Stenner et al. 2010; Foster and Walker 2014). Policy initiatives need to make sure that environments are created so as to make healthy choices available and easy for all individuals alike. Nevertheless, precarity may generate health inequalities among aging individuals. There are several pathways that can lead to the accumulation of health inequalities over the life span resulting in worse health in old age. One path involves not engaging in preparatory activities for old age such as regular health check-ups or leading a healthy lifestyle. A more indirect pathway involves the effect of negative views on aging on health and well-being across the life span. A third way in which precarity may determine health inequalities in old age is through the detrimental health effects of precarious work itself. In the following, I will explore these potential pathways in which precarity can foster the accumulation of health inequalities over the adult life span.

Health inequalities might emerge between precarious and financially secure individuals because the former prepare less for old age in comparison to the latter. Both groups perceive the uncertainty of the pension system as well as the decline inherent to the aging body. However, financially secure individuals can rely on their financial assets and take concrete actions to prepare for old age. Thus, even if there is no 100% certainty, at least having some form of financial security for old age allows for more positive views on aging.

Engaging in health behavior constitutes one way of preparing for old age. Adopting a healthier lifestyle and training body as well as mind were strategies mentioned by financially secure individuals in the interviews. In accordance with active aging policy principles, financially secure individuals consider health a matter of personal responsibility. For financially secure people, health constituted a matter of decision making. If one was sufficiently motivated, one could change ones' lifestyle. For financially secure individuals, there is a clear link between agency and health behavior. However, for precarious people even when they recognized the relevance of a healthy lifestyle, they perceived they lacked the necessary resources to help them achieve their health goals. Precarity represented a barrier toward adopting healthier lifestyles. Because of their living and work conditions precarious workers often cannot adopt the health behavior they would like to. For example, the precarious participants felt they lacked time for practicing sports and considered a healthy diet to be expensive.

Precarity offers a limited set of opportunities for health and thus can affect individual choices and lifestyles in a negative way. Katz and Calasanti (2015) affirm that the individual lifestyle and volition fit in the contemporary consumerist thought pattern that dominates health and retirement policy. Findings from the interviews reflect such thinking patters, where financially secure individuals affirm that they have started to train their body and mind in order to ensure an active old age. Precarious individuals have also internalized the discourse and consider healthy changes important for a good old age but are overwhelmed by the obstacles of precarity when they try to make lifestyle choices.

According to previous criticism of the successful aging theory, material conditions constrain individual volition and lifestyle choices (Katz and Calasanti 2015; Hendricks and Hatch 2009). People with access to health benefits and services also can frame their health behaviors to achieve positive outcomes (Katz and Calasanti 2015). Critics of the successful aging model conclude that lifestyle, choice, health, and positive aging are concepts that need to be theorized beyond personal choice because lifestyles are configured by different opportunities (Calasanti and King 2011; Dannefer 2006).

In conformity with previous critics of the successful aging model (Martinson and Berridge 2015; Liang and Luo 2012; Minkler and Holstein 2008; Katz and Marshall 2003), the active aging policy creates a positive aging standard of success that is unattainable for precarious individuals. Thus, inequalities in old age could emerge because the positive aging standards act as a barrier for precarious people, while for financially secure individuals they constitute a resource, a motivating factor for their preparatory efforts.

Having positive views on aging is a good predictor for health and well-being in later years (Wurm et al. 2010, 2013) and thus constitute a relevant resource when preparing for old age. In contrast to financially secure individuals, precarious people perceive more negative views on aging and more barriers toward a positive old age. As I have discussed in previous chapters, precarity can foster the internalization of negative images about aging. Negative self-perceptions of aging erode people's self-confidence that they could reach positive aging standards dictated by the active aging policy framework. Furthermore, previous studies suggested that negative aging views tend to become self-fulfilling prophecies (Wurm et al. 2013). Thus, the gap between positive aging standards and reality is enhanced and can lead to accumulating health disadvantages for old age.

According to the stereotype embodiment theory (Levy 2009), individuals assimilate stereotypes of aging from their surrounding culture and turn them into self-definitions of aging. These self-perceptions of aging influence the individuals' health and behavior (Levy 2009). Several studies have shown that views on aging or self-perceptions of aging affect health (Wolff et al. 2014; Levy 2009; Wurm et al. 2010, 2007). Different pathways were used as explanations. The physiological processes (Levy et al. 2000) and psychological mechanisms (Levy et al. 2002) and the behavioral path (Wolff et al. 2014; Levy and Myers 2004). Because precarious individuals cannot implement their healthy aging goals, their negative images of old age may become self-fulfilling prophecies (Wurm et al. 2013). Therefore, precarious workers could become the old people that they dread to become, namely ill and dependent.

The responsibility for one's health and adopting a healthy lifestyle is more difficult for precarious individuals. Precarious work conditions have been shown to be linked to bad physical (Lewchuk et al. 2008; Kim and von dem Knesebeck 2015; László et al. 2010) and mental health (Han et al. 2017; Canivet et al. 2017; Min et al. 2015; Moscone et al. 2016). Findings from the interviews also point out that precarious individuals experience barriers in maintaining their physical and mental health. One obstacle concerns the material costs of the healthy lifestyle (e.g., healthy food is more expensive than junk food) and the time barrier. World Health Organization recommendations include practicing moderate physical activity and overall leading a healthy life (e.g., healthy eating, not smoking, etc.). However, precarious individuals have full schedules that leave little time for practicing sports or integrating healthy routines in their lives. Risk factors for precarious people include the constant anxiety and stress that they feel regarding their work and the lack of time they experience in what concerns leisure activities. One aspect that worried almost all precarious study participants was the cost of falling ill. If precarious individuals are not able to work, there is no insurance to cover the costs of living and they run a high risk of becoming dependent in their old age. Since these health disadvantages accumulate over the years, precarious individuals risk an old age characterized by illness and disability.

There is a whole body of literature that links precarious work to health problems (Ferrie 2002; Ferrie et al. 2005; Benach et al. 2002). Precarious work is defined through dimensions such as temporality, powerlessness, and lack of social benefits (Benach and Muntaner 2007). The reduced job security, lower benefits (e.g., lower payment, no health insurance, no vacations, etc.), and impaired working conditions all increase the risk for physical and mental illnesses in precarious workers (Moscone et al. 2016; Kim and von dem Knesebeck 2015; Kim et al. 2012; Lewchuk et al. 2008; Ferrie et al. 2005; Benach et al. 2002). For Germany in particular, it was shown that precarious workers or the "working poor" have worse health compared to the population in general (Broding et al. 2010).

Job insecurity in particular is associated with chronic anxiety, and it was shown that the negative effects of the stressor linger even after the removal of the threat, for example, when precarious workers get a stable job (Ferrie 2002). Precarious work was shown to be associated with high job stress (Kachi et al. 2014) and negative effects on mental health (Canivet et al. 2017). Another study found high associations between precarious work and depression in both women and men (Ferrie et al. 2005). Non-permanent workers enjoy less autonomy at work and less control over working time than permanent employees (Benach et al. 2002). Especially the combination between low job control and high job demands is particularly detrimental for physical health (Landsbergis et al. 2003). Women working in precarious jobs have particularly high risk for health problems (Ferrie et al. 2005) and depression (Kim et al. 2012). Kim and von dem Knesebeck (2015) compared the effects of unemployment and insecure jobs on health and found that the negative health consequences are very similar. Job insecurity was highly related with somatic symptoms, while unemployment was related to worse health in general (Kim and von dem Knesebeck 2015). Studies that focused on mental health and precarious work showed that insecure employment can lead to a high suicide ideation and attempts (Min et al. 2015) especially in male workers (Han et al. 2017).

All in all, the findings lend support to the existing evidence base concerning the negative effects of precarious work on health (Ferrie 2002; Ferrie et al. 2005; Benach et al. 2002). Thus, working in a precarious job can lead to the accumulation of health problems over the life span and health inequalities in old age. Precarity shapes the environment in which people are aging. Existential vulnerabilities are experienced across the life span and determine the accumulation of disadvantages that would result in a more negative old age. Precarious individuals lack resources and are more vulnerable when facing aging in comparison with financially secure individuals. The vulnerability results from the frailty of the body as well as from the insecure work and living conditions.

Ageism at Work

Older workers in general face difficulties when trying to find employment in their late life and are also more prone to lose their jobs to younger individuals (Johnson 2009; Roscigno 2010). Unfortunately, ageism is still a reality on the job market, even if efforts are made to employ older individuals and to postpone retirement age (Foster and Walker 2014).

The potential for discrimination on the job market is generally higher for precarious individuals, and it gets more difficult to find a job as one gets older. Financially secure people have the choice to work after retirement or engage in voluntary actions. People who have worked in precarious jobs need to struggle to keep these jobs when they are older. Furthermore, when they apply for jobs, they are competing against other precarious workers of all ages as well as financially secure individuals who may want to work part time after retirement. Precarious individuals are more exposed to ageist discrimination as compared to financially secure individuals. Moreover, their capacity for agency is limited by the uncertainty of work conditions. Precarious individuals face a double risk for discrimination: first because they are old or getting older and second because they are working in precarious jobs that place them at the margins of the job market. Concerning the latter, for instance a patchwork CV may place a person outside the competition for a desirable stable post. Moreover, a CV with precarious jobs may not seem "stable" or "secure" enough for potential employers. The likelihood that precarious workers would become the target of human resources head hunters is very small.

Results from the interviews show that precarious individuals feel threatened by younger people in what concerns working in old age. They have to compete for the precarious jobs with younger precarious workers but also with financially secure individuals who want to work part time after retirement. Such competition for the limited work positions can only harm intergenerational communication as well as relations with aged peers. Negative social relations with peers and younger generations can only emphasize the precarity of the experienced social reality. As discussed in Chap. 4, precarity can determine a feeling of disaffiliation (Castel 2000). Insecure social relations make people feel vulnerable at social and emotional levels (Castel 2000). As a consequence of disaffiliation, precarious individuals may feel lonely and isolate themselves from society and thus further reduce their chances of finding better, secure jobs.

The active aging policy encourages not only work in old age, but volunteer activities for older adults which were proven to be beneficial for their health and well-being (Lum and Lightfoot 2005; Greenfield and Marks 2004; Piliavin and Siegl 2007). Volunteer work constitutes a good choice for financially secure individuals, but less so for precarious workers who have to make a living in old age. Findings from the interviews show that when there is a retirement to rely on, people have already started taking concrete actions toward volunteer activities. In contrast, precarious workers do not think they would have time for volunteering in old age, although some of them expressed the desire to do so. Volunteering can represent a meaningful activity in old age, an opportunity for civic engagement with positive benefits for the community and individuals alike (Konrath et al. 2012; Lum and Lightfoot 2005; Morrow-Howell et al. 2003). However, precarious circumstances can represent a barrier toward engaging in volunteering activities. People who work in precarious jobs are less likely to engage in volunteer activities, since they need to make a living and have little time left except for work and job seeking. Thus, an important resource such as volunteering is not easily accessible to precarious individuals.

All in all, by not working in old age or not volunteering, precarious individuals become part of the negative stereotype about older inactive individuals. They may become "a burden" or "a threat" to the social order and thus justify further ageist attitudes on the job market or society in general.

Mental Barriers in Precarious Times

Precarity shapes preparations for old age by fostering certain mental barriers that hinder preparatory efforts. These mental barriers include a feeling of lacking control, low self-confidence, and the perception that making plans for the future is pointless. In the following, I will analyze each of these and how they affect how precarious workers prepare or not prepare for old age.

Lack of Control

Precarity fosters a feeling of lack of control over time (especially present and future) and over ones' life in general (e.g., work, work and life balance, health, etc.). Although in some cases it might be that the feeling of lacking control is subjective, in most cases, the objective control may also be missing because of the precarious circumstances.

Findings from the interviews confirm the fact that the precarious middle-aged Germans experienced little control over their time. For instance, concerning their present, they felt they cannot take time off from work to relax and they have "no time" for hobbies or leisure activities. In this sense, they also considered they had "not enough time" to prepare for old age and did not consider such preparations a priority. The latter is linked to the feeling of "no future" (Foti 2017) and lack of control over an insecure future. Moreover, the feeling of lacking control makes precarious individuals avoid planning their old age. This avoidance is connected to the fear that they cannot make their plans become reality in an unreliable future. Precarity makes achieving goals in old age seem uncertain and generates feelings of anxiety. In this sense, some interviewed precarious individuals said they prefer not to plan as a way to prevent feeling anxious about old age. By contrast, financially secure individuals feel anxious about not planning their old age.

Precarious work has been defined through dimensions such as a feeling of powerlessness, lack of benefits, and temporality (Kalleberg 2009; Benach and Muntaner 2007). These distinctive dimensions imply that a person feels no control over the type of job they do, how much they are paid, and how long they will be able to hold on to a particular job. Precarious workers feel an overwhelming pressure to be constantly active, always searching for a new job or novel hiring opportunity. This state of constant alertness makes concentration on formulating goals difficult. Precarious individuals experience a high amount of stress and anxiety in their daily lives that can interfere with setting goals and feeling in control to achieve them. The pressure to be active grows with age, as people realize they should make plans for retirement, in addition to finding work at present. It is to be expected that uncertainty and vulnerability would trigger anxiety and block people from action. Thus, precarity interferes with preparatory efforts and raises the chances of developing inequalities in old age. Both financially secure and precarious individuals are affected by the ethos of productivity, usefulness, and activity. Nevertheless, precarious workers feel a more acute dissonance between their precarious reality and the pressure to age well. They are aware of their anxiety and fight it by not planning and by trying to frame the uncertainties of their daily life in a positive way.

Lack of Confidence: The Precarious Identity

Precarious individuals want to stay independent, active, and healthy in old age. In this sense, they embrace the ideal of positive aging. However, what they lack is the confidence and agency to achieve these goals. Since precarity inspires little self-efficacy in middle-aged individuals, even if the opportunities for agency in old age arise, people have little confidence that they can take advantage of them. In general, precarious individuals show a lack of confidence in themselves and their future.

In a culture where people define themselves through what they do (e.g., what kind of work, what kind of hobbies and activities they choose), being out of work or having no time for hobbies put the "self" in danger. Volunteer activities can provide a sense of meaning to old age (Konrath et al. 2012). After retirement, volunteer work offers the opportunity to do something for others, for the community but also for oneself. Volunteer actions structure the free time one has after retiring from work. For example, financially secure individuals have mentioned preparing to engage in volunteer work after retirement. Even among the precarious individuals, a few also already engaged in volunteer work. Volunteering helps them structure their time and prevent the negative effects of unemployment. Moreover, volunteering provides access to a social network, social support, and the feeling of being socially engaged. Nevertheless, for precarious individuals, volunteering is not something they can plan to do in their old age since they would need to engage in paid work in order to ensure their well-being and living expenses. In this sense, precarity makes volunteering more difficult in old age and thus takes away an important asset from precarious workers.

Autonomy represents an important goal for both precarious and financially secure individuals. Nevertheless, financially secure individuals are proactive about how to change their environment in order to achieve positive aging, while precarious individuals focus more on adaptive coping. For example, they use positive reframing as a mechanism to make living in a precarious reality seem easier. However, by adjusting themselves to the precarious reality, their sense of self is affected. Their self-identity may suffer from being too accustomed to living in precarious times (e.g., they become too flexible, cannot make plans, are afraid to pursue their hobbies due to lack of time, etc.). This adaptation also stops them from actions that could change their current situation.

Both precarious and financially secure individuals have internalized the neoliberal ideology that holds them responsible for their own health and financial situation in old age (Katz and Marshall 2003; Minkler and Holstein 2008). If they are not able to achieve the successful aging ideal, they are at risk of feeling bad about themselves. As a result, their self-efficacy would decrease. Being a precarious old person represents a "dangerous identity" because it is linked with negative connotations such as poverty or failure, discrimination, of "*not being on the top of the evaluation scale*" as one of the precarious participants described this feeling.

In chapter five, this feeling was described by precarious individuals as being a "mental void," a state of mind of uncertainty, where one feels vulnerable and lost because they perceive an acute insecurity and a feeling of "no future." Similar results were found by studies that have shown an association between precarious work and depression or suicidal ideation (Han et al. 2017; Kim et al. 2012).

Preparatory efforts concerning old age are quite different between precarious and financially secure individuals. While the first do not see the point in making arrangements for old age or consider planning to be difficult, the later carefully plan their retirement years. This confirms the fact that positive aging policy is creating a standard that only financially secure individuals can reach (Katz and Marshall 2003; Martinson and Berridge 2015; Katz and Calasanti 2015). People who can rely on their pensions can envisage a future where they can be socially engaged (e.g., through volunteer activities) or care about their health and well-being. While the future fills financially secure individuals with hope and encourages them to plan, for precarious individuals, planning is generating anxiety or feelings of helplessness toward reaching their desired old age. Planning in case of precarious individuals reminds them of the gap between their precarious reality and the promoted positive aging ideal.

Reinventing Preparation for Old Age

According to active aging recommendations, preparation for a good old age should constitute a lifelong project (see successful aging theories and policies, Chap. 2 and 3) accessible to all individuals alike. However, middle-aged precarious workers are disadvantaged in what preparation according to policy principles is concerned. Findings point out that German middle-aged individuals working in precarious jobs were" surprised" by the necessity to organize some form of financial preparation for their old age. Precarious middle-aged people felt they have started late in life to prepare for old age and are therefore at a disadvantage in comparison with financially secure people. Younger generations of precarious people may have it easier since they "know what to expect" to a certain extent and might thus prevent the accumulating disadvantages.

One domain where inequalities can accumulate is health, as I have discussed above. Health constitutes a relevant asset for old age for both precarious and financially secure individuals. Health inequalities may appear in old age as a result of the investment/non-investment in the healthy body. Most of the interviewees mentioned that they are doing regular medical check-ups. However, in what concerns engagement in health behavior, precarious individuals mentioned several barriers like lacking control over time or expenses related to health. Short-term thinking constituted another barrier for precarious individuals. Since they are concerned with the present and not the future, they do not engage in health behavior that may have positive outcomes in the long run. They also work long hours and are constantly under pressure to perform and get new jobs. In contrast, financially secure people focus on the future and the things they will be able to enjoy when they retire. Financially secure individuals value discipline and the only barrier they encountered was lacking this discipline for regular engagement in health behavior. Nonetheless, investing in health is not the only kind of preparation one can engage in for old age. As a fact, literature concerning preparations for old age was criticized for focusing too much on health issues (i.e., healthism) and ignoring other modalities of preparation. Furthermore, it focused too much on personal responsibility toward a healthy old age and too little on reducing barriers concerning the actions required for aging well.

The active aging policy insists on promoting the agency of older people (Jolanki 2009). In the "new aging" discourse (Katz and Calasanti 2015), older individuals are required to be self-sufficient and take responsibility for their own lives. In the process of aging, individual choices and actions become very important. Thus, individuals are expected to design their own old age. For financially secure individuals, this challenge represents something to look forward to. Retirement promises opportunities of self-realization, involvement in social activities, release from social obligations, the possibility to be an active member of society (Jolanki 2009). However, precarious jobs do not provide the security of retirement. For precarious individuals, the idea of staying productive is not constructed as an opportunity, but a requirement in order to make a living in old age.

Making choices is difficult in the context of precarity. True agency can exist when people have the self-efficacy to make changes in themselves and their social environment (Jolanki 2009). Apart from the availability of choices, the ability to make and enact these choices needs to be present (Jolanki 2009; Morgan et al. 2006). Financially secure individuals are convinced that they can influence their aging process through their own actions. Thus, they invest time and energy in training their body and their mind to keep fit in old age. Contrasting with these "classical" measures (i.e., in accordance with solutions proposed by active aging policy), precarious individuals find original ways to enact their agency. For instance, they train their work-related skills, invest time in social resources such as building social networks, take a holiday that is connected to a work project, or spend quality time with family or friends. In order to prepare for old age and have a feeling of their own agency, precarious individuals reinvent preparation. This means that they take the active aging model and redesign it to fit their social reality. For example, they constantly reinvent themselves to stay eligible on the job market, they build social networks and "train" themselves to remain optimistic. Furthermore, they rely more on the accumulation of cognitive and social resources and their effective management as preparation for old age as on planning. For precarious individuals, preparation for old age is an ongoing, flexible, and dynamic process rather than a structured and carefully planned project.

As discussed in the previous chapters, precarity as a social context creates more barriers for positive aging than it provides resources. Thus, in order to prepare for old age in precarious contexts, middle-aged individuals have to think differently about preparation. The German society in particular is focused on a positive discourse of aging and on promoting active aging policy guidelines. However, such recommendations work better for financially secure individuals. Interview findings show that financially secure individuals internalize the active aging discourse and apply it like a "to do" list in their preparations for old age. For precarious workers, preparation measures need to be more flexible and applied in a flexible way as precarity requires. Sometimes what seems not to be a preparatory action represents an effective strategy. Even if precarious individuals desire a good old age, they experience many barriers in achieving such goals. Thus, in order to prepare for old age, precarious middle-aged individuals have to find novel ways to prepare as well as identify resources that would facilitate positive aging. During the interviews, precarious middle-aged Germans emphasized the importance of cognitive and social aspects concerning the preparation for old age.

All in all, in order to prevent the accumulation of disadvantages (e.g., concerning health or financial assets), precarious workers need to reinvent preparation for old age.

The main strategies that they apply are not planning as a sort of anti-strategy, engaging in positive reframing and social networking to maintain their social relations.

As I have discussed throughout the book, planning constitutes an important ingredient when it comes to preparation for old age within the active aging framework (Foster and Walker 2014; Walker 2009). People are supposed to plan what they will do after retirement and find the means to achieve these goals. However, findings from the interviews show that precarious individuals do not want to plan their old age. Planning requires imagining a long-term perspective, which is hard for precarious workers and thus triggers negative emotions. Moreover, planning generates anxiety since precarious individuals are not confident that they have the means to reach their goals. On a psychological level, "not planning" becomes a coping mechanism when facing the uncertainties of old age. Precarious individuals prefer to think short term and concentrate on what they can achieve here and now.

For precarious middle-aged Germans, the challenge is how they deal with **the cognitive dissonance** between the positive image that is presented in the media (i.e., active, working old individual) and their negative images of old age (poor, ill old people) which are based on fears emerging from their precarious lives. The **cognitive dissonance between positive and negative images** of aging is experienced by precarious individuals, and they seek ways to deal with this dissonance. One strategy applied by precarious individuals is to think that they should "expect nothing from old age." When you have no expectations, you are either positively surprised or you cannot be disappointed because you did not have expectations in the first place. A further strategy is to focus on the positive image of old age and try to live up to it. However, this positive aging ideal can become a burden because they feel a pressure to live up to this expectation. At the same time, they may experience an acute fear of failure since precarious living conditions are less likely to be associated with a positive old age. All in all, not planning represents a strategy to cope with negative emotions generated by precarity (e.g., anxiety, anger) and to solve the cognitive dissonance between active aging ideals and the precarious reality.

Precarious workers believe that if you keep a youthful attitude to life, you manage to stay young. Thus, they have internalized the "eternal youth ideal" since they need to stay young to get jobs in old age. It also implies they have internalized the negative stereotype that old people are less valuable on the job market because they are less healthy and less clear minded. In terms of ***cognitive*** preparatory efforts, both precarious and financially secure individuals thrive to keep an active mind. Interviewees from both groups mentioned how "they trained their mind" consciously through reading or mental exercises. Precarious individuals did this also by trying to keep a positive perspective, being open to new ideas.

As I discuss in the previous chapter, precarious individuals do hold positive views on aging that resemble the positive images of aging promoted by mass media (i.e., athletic old people running or happy grandparents playing with their grandchildren). The positive image of old age is of an active, healthy, young looking individual who keeps working and continues to be useful to society and enjoy life. This image corresponds to the one presented by policies such as active aging or positive aging (Lassen and Moreira 2014; Foster and Walker 2014; Walker 2009) and sometimes promoted by mass media (Rozanova 2010). Participants mentioned that they expect to become more relaxed with old age in what dealing with negative emotions is concerned. However, when they talked about their personal indicators of old age, these pointed to loss and illness (i.e., negative view on aging) and not to growth and development (i.e., positive view on aging). The problem with negative images of aging is that they can become self-fulfilling prophecies (see Wurm et al. 2013). Thus, precarious individuals who associate aging with illness and decline are more prone to experience bad health in old age. They are also more likely to limit their activities as they approach a later stage in their lives since they might feel their resources are depleting. A precarious person may invest less energy in activities that are pleasant and invest all energy in work since the latter is important for survival in the moment as well as in old age.

Precarious individuals take care to maintain and enlarge their social networks. For them, social relations are crucial for work opportunities and also for emotional comfort. For the former, meeting new people is relevant, while for the latter maintaining good relations with friends and family is vital. Some financially secure people also plan to increase their social networks, for instance those who plan to engage socially, who want to do something for the community. However, for them, expanding social networks is stimulated by wanting to stay youthful and avoid appearing old and rigid.

Findings also lend support to the self-regulation theory (De Ridder and de Wit 2006) where optimism and goal setting are related. However, while financially secure individuals focus on gain-related goals (e.g., train body and mind), precarious people focus on loss-avoidance goals (e g. avoid loss of skills needed at work, avoid losing social networks). The latter are confident they can maintain their social relations and effectively deal with losses that come with aging. Aging itself constitutes a process where people gradually exercise less primary control (i.e., change circumstances according to personal goals) and more secondary control (i.e., adjust personal goals to make them consistent with situational constraints). Precarity implies the application of secondary control since it involves changing positive aging goals to adjust to situational limitations. According to socioemotional selectivity theory (Carstensen 2006), with increasing age, people perceive future time left in life as limited, and prioritize emotional goals. Nevertheless, present findings show that precarious workers favor both emotional and instrumental goals despite perceiving a limited time perspective.

Some assets for old age are similar to financially secure individuals and also to what the active aging policy recommends. However, other resources need to be redesigned. If you cannot rely on the State, you go back to the available social resources such as family and friends. Social relations were found in previous literature to be considered resources for old age (Jopp et al. 2014; Hilton et al. 2012; Jolanki 2009).

Precarious individuals rely on resources for old age such as building social networks, spending quality time with life partners or family. Compared to them, financially secure individuals want to be more independent from their families. Moreover, they concentrate on becoming social activists in old age because they want to do something for the community. Some of the interviewed participants already engage in voluntary activities while others plan to do so in old age. The motives for volunteering were shown to be important for health. This means that not all forms of volunteering are beneficial for health. Findings pointed out that only volunteering for other-oriented reasons had a decreased mortality risk (Konrath et al. 2012). In this sense, precarious individuals do volunteer for other-oriented reasons and may benefit from the health effects of volunteering.

All in all, both precarious and financially secure persons share the same precarious reality. This becomes evident in the fact that financially secure persons also do not trust the pension system. Yet, the difference lies in the fact that they ensure they have alternative resources such as real estate properties or private pensions, while precarious individuals only talk about how they would look for such alternatives if they could afford these. Nonetheless, precarious workers create their own resources for old age, showing that positive aging constitutes an ideal for them as well. Precarity can constitute a resource that enhances their creativity for refining their survival skills. In this line of argument, within the next chapter, I explore the question of whether positive aging is possible in precarious times.

References

Benach, J. M., Amable, C. Muntaner, & Benavides, F. G. (2002). The consequences of flexible work for health: are we looking at the right place? *Journal of Epidemiology and Community Health, 56*(6), 405–406. https://doi.org/10.1136/jech.56.6.405.

Benach, J., & Muntaner, C. (2007). Precarious employment and health: Developing a research agenda. *Journal of Epidemiology and Community Health, 61*(4), 276–277. https://doi.org/10.1136/jech.2005.045237.

Biggs, S. (2014). Precarious ageing versus the policy of indifference: International trends and the G20. *Australasian Journal on Ageing, 33*(4), 226–228.

Breheny, M., & Stephens, C. (2010). Ageing in a material world. *New Zealand Journal of Psychology, 39*(2), 41–48.

Broding, H. C., Weber, A., Glatz, A., & Bünger, J. (2010). Working poor in Germany: Dimensions of the problem and repercussions for the health-care system. *Journal of Public Health Policy, 31*(3), 298–311. https://doi.org/10.1057/jphp.2010.20.

Calasanti, T., & King, N. (2011). A feminist lens on the third age: Refining the framework. In D. Carr & K. Komp (Eds.), *Gerontology in the era of the third age* (pp. 67–85). New York: Springer.

Canivet, C., Aronsson, G., Bernhard-Oettel, C., Leineweber, C., Moghaddassi, M., Stengård, J., … Östergren, P.-O. (2017). The negative effects on mental health of being in a non-desired occupation in an increasingly precarious labour market. *SSM - Population Health, 3,* 516–524. https://doi.org/10.1016/j.ssmph.2017.05.009.

Carstensen, L. L. (2006). The influence of a sense of time on human development. *Science, 312*(5782), 1913–1915. https://doi.org/10.1126/science.1127488.

Castel, R. (2000). The roads to disaffiliation: Insecure work and vulnerable relationships. *International Journal of Urban and Regional Research, 24*(3), 519–535.

Craciun, C., & Flick, U. (2015). "I want to be 100 years old, but I smoke too much": Exploring the gap between positive aging goals and reported preparatory actions in different social circumstances. *Journal of Aging Studies, 35,* 49–54. https://doi.org/10.1016/j.jaging.2015.07.006.

Craciun, C., & Flick, U. (2014). "I will never be the granny with rosy cheeks": Perceptions of aging in precarious and financially secure middle-aged Germans. *Journal of Aging Studies, 29,* 78–87. https://doi.org/10.1016/j.jaging.2014.01.003.

Dannefer, D. (2006). Reciprocal co-optation: The relationship of critical theory and social gerontology. In J. Baars, D. Dannefer, C. Phillipson, & A. Walker (Eds.), *Aging, globalization and inequality: The new critical gerontology* (pp. 103–120). Amityville, NY: Baywood.

De Ridder, D. T. D., & de Wit, J. B. F. (Eds.). (2006). *Self-Regulation in Health Behavior*. John Wiley and Sons. https://doi.org/10.1002/9780470713150.

Ferrie, J. E., Shipley, M. J., Newman, K., Stansfeld, S. A., & Marmot, M. (2005). Self-reported job insecurity and health in the Whitehall II study: Potential explanations of the relationship. *Social Science and Medicine, 60*(7), 1593–1602. https://doi.org/10.1016/j.socscimed.2004.08.006.

Ferrie, J. E. (2002). Effects of chronic job insecurity and change in job security on self reported health, minor psychiatric morbidity, physiological measures, and health related behaviours in British civil servants: The Whitehall II study. *Journal of Epidemiology and Community Health, 56*(6), 450–454. https://doi.org/10.1136/jech.56.6.450.

Foster, L., & Walker, A. (2014). Active and Successful Aging: A European Policy Perspective. *The Gerontologist, 55*(1), 83–90. https://doi.org/10.1093/geront/gnu028.

Foti, A. (2017). *General Theory of the Precariat. Great Recession, Revolution, Reaction.* Downloaded from http://networkcultures.org/publications.

Fung, H. H. (2013). Aging in Culture. *The Gerontologist, 53*(3), 369–377. https://doi.org/10.1093/geront/gnt024.

Grenier, A., Phillipson, C., Laliberte Rudman, D., Hatzifilalithis, S., Kobayashi, K., & Marier, P. (2017). Precarity in late life: Understanding new forms of risk and insecurity. *Journal of Aging Studies, 43,* 9–14. https://doi.org/10.1016/j.jaging.2017.08.002.

Greenfield, E. A., & Marks, N. F. (2004). Formal volunteering as a protective factor for older adults' psychological well-being. *The Journals of Gerontology Series B: Psychological Sciences and Social Sciences, 59*(5), 258–264.

Johnson, R. W. (2009). The recession's impact on older workers. *Public Policy and Aging Report, 19,* 26–31.

Jolanki, O. H. (2009). Agency in talk about old age and health. *Journal of Aging Studies, 23*(4), 215–226. https://doi.org/10.1016/j.jaging.2007.12.020.

Jopp, D. S., Wozniak, D., Damarin, A. K., De Feo, M., Jung, S., & Jeswani, S. (2014). How could lay perspectives on successful aging complement scientific theory? Findings from a U.S. and a German Life-Span sample. *The Gerontologist, 55*(1), 91–106. https://doi.org/10.1093/geront/gnu059.

Han, K.-M., Chang, J., Won, E., Lee, M.-S., & Ham, B.-J. (2017). Precarious employment associated with depressive symptoms and suicidal ideation in adult wage workers. *Journal of Affective Disorders, 218,* 201–209. https://doi.org/10.1016/j.jad.2017.04.049.

Hendricks, J., & Hatch, L. R. (2009). Theorizing lifestyle: Exploring agency and structure in the life course. In V. L. Bengston, D. Gans, N. M. Putney, & M. Silverstein (Eds.), *Handbook of theories of aging* (pp. 435–454). New York: Springer.

Hilton, J. M., Gonzalez, C. A., Saleh, M., Maitoza, R., & Anngela-Cole, L. (2012). Perceptions of successful aging among older Latinos, cross-cultural context. *Journal of Cross-Cultural Gerontology, 27*(3), 183–199. https://doi.org/10.1007/s10823-012-9171-4.

Kachi, Y., Otsuka, T., & Kawada, T. (2014). Precarious employment and the risk of serious psychological distress: A population-based cohort study in Japan. *Scandinavian Journal of Work, Environment & Health, 40*(5), 465–472. https://doi.org/10.5271/sjweh.3442.

Kalleberg, A. L. (2009). Precarious Work, Insecure Workers: Employment Relations in Transition. *American Sociological Review, 74*(1), 1–22. https://doi.org/10.1177/000312240907400101.

Katz, S., & Marshall, B. (2003). New sex for old: Lifestyle, consumerism, and the ethics of aging well. *Journal of Aging Studies, 17*(1), 3–16.

Katz, S., & Calasanti, T. (2015). Critical Perspectives on Successful Aging: Does It "Appeal More Than It Illuminates"? *The Gerontologist, 55*(1), 26–33. https://doi.org/10.1093/geront/gnu027.

Kim, I. H., Muntaner, C., Vahid Shahidi, F., Vives, A., Vanroelen, C., & Benach, J. (2012). Welfare states, flexible employment, and health: A critical review. *Health Policy, 104*(2), 99–127. http://dx.doi.org/10.1016/j.healthpol.2011.11.002.

Kim, T. J., & von dem Knesebeck, O. (2015). Is an insecure job better for health than having no job at all? A systematic review of studies investigating the health-related risks of both job insecurity and unemployment. *BMC Public Health, 15*(1). https://doi.org/10.1186/s12889-015-2313-1.

Konrath, S., Fuhrel-Forbis, A., Lou, A., & Brown, S. (2012). Motives for volunteering are associated with mortality risk in older adults. *Health Psychology, 31*(1), 87–96. https://doi.org/10.1037/a0025226.

Lassen, A. J., & Moreira, T. (2014). Unmaking old age: Political and cognitive formats of active ageing. *Journal of Aging Studies, 30*, 33–46. https://doi.org/10.1016/j.jaging.2014.03.004.

Landsbergis, P. A., Schnall, P. L., Pickering, T. G., Warren, K., & Schwartz, J. E. (2003). Lower socioeconomic status among men in relation to the association between job strain and blood pressure. *Scandinavian Journal of Work, Environment & Health, 29*(3), 206–215.

László, K. D., Pikhart, H., Kopp, M. S., Bobak, M., Pajak, A., Malyutina, S., … Marmot, M. (2010). Job insecurity and health: A study of 16 European countries. *Social Science and Medicine, 70*(6), 867–874. https://doi.org/10.1016/j.socscimed.2009.11.022.

Levy, B. (2009). Stereotype Embodiment. *Current Directions in Psychological Science, 18*(6), 332–336. https://doi.org/10.1111/j.1467-8721.2009.01662.x.

Levy, B. R.; Slade, M. D., Kasl, S. V. (2002): Longitudinal benefit of positive self-perceptions of aging on functional health. *Journal of Gerontology, 57*(5), 409–417.

Levy, B. R., & Myers, L. M. (2004). Preventive health behaviors influenced by self-perceptions of aging. *Preventive Medicine, 39*, 625–629.

Levy, B. R., Hausdorff, J. M., Hencke, R., & Wei, J. Y. (2000). Reducing cardiovascular stress with positive self-stereotypes of aging. *The Journals of Gerontology Series B: Psychological Sciences and Social Sciences, 55*(4), P205–P213. https://doi.org/10.1093/geronb/55.4.p205.

Lewchuk, W., Clarke, M., & de Wolff, A. (2008). Working without commitments: Precarious employment and health. *Work, Employment & Society, 22*(3), 387–406. https://doi.org/10.1177/0950017008093477.

Liang, J., & Luo, B. (2012). Toward a discourse shift in social gerontology: From successful aging to harmonious aging. *Journal of Aging Studies, 26*(3), 327–334.

Lum, T. Y., & Lightfoot, E. (2005). The effects of volunteering on the physical and mental health of older people. *Research on Aging, 27*, 31–55. https://doi.org/10.1177/0164027504271349.

Martinson, M., & Berridge, C. (2015). Successful aging and its discontents: A systematic review of the social gerontology literature. *The Gerontologist, 55*(1), 58–69. https://doi.org/10.1093/geront/gnu037.

Minkler, M., & Holstein, M. B. (2008). From civil rights to civic engagement? Concerns of two older critical gerontologists about a "new social movement" and what it portends. *Journal of Aging Studies, 22*(2), 196–204. https://doi.org/10.1016/j.jaging.2007.12.003.

Min, K.-B., Park, S.-G., Hwang, S. H., & Min, J.-Y. (2015). Precarious employment and the risk of suicidal ideation and suicide attempts. *Preventive Medicine, 71*, 72–76. https://doi.org/10.1016/j.ypmed.2014.12.017.

Morrow-Howell, N., Hinterlong, J., Rozario, P. A., & Tang, F. (2003). Effects of volunteering on the well-being of older adults. *The Journals of Gerontology: Series B, 58*(3), S137–S145. https://doi.org/10.1093/geronb/58.3.s137.

Morgan, L. A., Eckert, J. K., Piggee, T., & Frankowski, A. C. (2006). Two lives in transition: Agency and context for assisted living residents. *Journal of Aging Studies, 20*, 123–132.

Moscone, F., Tosetti, E., & Vittadini, G. (2016). The impact of precarious employment on mental health: The case of Italy. *Social Science and Medicine, 158*, 86–95. https://doi.org/10.1016/j.socscimed.2016.03.008.

Moulaert, T., & Paris, M. (2013). Social policy on ageing: The case of "Active Ageing" as a theatrical metaphor. *International Journal of Social Science Studies, 1*(2). https://doi.org/10.11114/ijsss.v1i2.141.

Piliavin, J. A., & Siegl, E. (2007). Health benefits of volunteering in the Wisconsin longitudinal study. *Journal of Health and Social Behavior, 48*(4), 450–464. https://doi.org/10.1177/002214650704800408.

Portacolone, E., Rubinstein, R. L., Covinsky, K. E., Halpern, J., & Johnson, J. K. (2018). The Precarity of older adults living alone with cognitive impairment. *The Gerontologist*. https://doi.org/10.1093/geront/gnx193.

Roscigno, V. J. (2010). Ageism in the American workplace. *Contexts, 9,* 16–21. https://doi.org/10.1525/ctx.2010.9.1.16.

Rozanova, J. (2010). Discourse of successful aging in The Globe and Mail: Insights from critical gerontology. *Journal of Aging Studies, 24*(4), 213–222. https://doi.org/10.1016/j.jaging.2010.05.001.

Standing, G. (2011). *The Precariat: The new dangerous class*. London: Bloomsbury Academic.

Stenner, P., McFarquhar, T., & Bowling, A. (2010). Older people and "active ageing": Subjective aspects of ageing actively. *Journal of Health Psychology, 16*(3), 467–477. https://doi.org/10.1177/1359105310384298.

WHO. (2002). *Active ageing: A policy framework*. Geneva: Author.

Walker, A. (2009). Commentary: The emergence and application of active aging in Europe. *Journal of Aging and Social Policy, 21*(1), 75–93. https://doi.org/10.1080/08959420802529986.

Wolff, J. K., Warner, L. M., Ziegelmann, J. P., Wurm, S., & Kliegel, M. (2016). Translating good intentions into physical activity: Older adults with low prospective memory ability profit from planning. *Journal of Behavioral Medicine, 39*(3), 472–482. https://doi.org/10.1007/s10865-015-9707-5.

Wolff, J. K., Warner, L. M., Ziegelmann, J. P., & Wurm, S. (2014). What do targeting positive views on ageing add to a physical activity intervention in older adults? Results from a randomised controlled trial. *Psychology and Health, 29,* 915–932. https://doi.org/10.1080/08870446.2014.896464.

Wurm, S., Tesch-Römer, C., & Tomasik, M. J. (2007). Longitudinal findings on aging-related cognitions, control beliefs, and health in later life. *The Journals of Gerontology, Series B: Psychological Sciences and Social Sciences, 62,* 156–164. https://doi.org/10.1093/geronb/62.3.P156.

Wurm, S., Tomasik, M. J., & Tesch-Römer, C. (2010). On the importance of a positive view on ageing for physical exercise among middle-aged and older adults: Cross-sectional and longitudinal findings. *Psychology and Health, 25*(1), 25–42. https://doi.org/10.1080/08870440802311314.

Wurm, S., Warner, L. M., Ziegelmann, J. P., Wolff, J. K., & Schüz, B. (2013). How do negative self-perceptions of aging become a self-fulfilling prophecy? *Psychology and Aging, 28*(4), 1088–1097. https://doi.org/10.1037/a0032845.

Chapter 10
Positive Aging in the Context of Precarity: Conclusions and Implications

Introduction

The question that makes the subject of this chapter and concludes the argument of the book is ***whether there can be positive aging in a context of precarity?*** Can the active aging policy framework (WHO 2002; Foster and Walker 2014) be reconciled with the precarity (Grenier et al. 2017; Grenier and Phillipson 2018) that nowadays characterizes the lives of middle-aged precarious workers? In order to answer this question, the chapter examines the thinking and acting patterns of precarious middle-aged Germans that were discussed throughout the book. Based on the different examples presented in the previous chapters, one can argue that positive aging can exist in precarious contexts, because precarity shapes the way people imagine old age and how they enact preparations for old age. In this new culture of risk, aging individuals find ways to survive precarity. However, this does not mean that they do not need help both at individual and policy levels in order to ensure health and well-being in old age. Therefore, in the present chapter I will discuss implications for psychological practice and policy in relation to positive aging and health. Finally, future research directions will be outlined in what concerns precarious aging.

Surviving Precarity and Aging Well

All throughout the book, the question has lingered, how can people survive precarity and age well? From the discussion about the experience of precarity in Chap. 5, one can see that middle-aged precarious individuals do not portray themselves as victims. Rather, they prefer to consider themselves "outside the system," as rebels, as people who are "skilled at surviving" in precarious times. Even if they perceive their own vulnerability and do live with many anxieties concerning their present and future, they adapt to the situation by internalizing thinking patterns that characterise precarity. These thinking patterns remind one of what has been called in the literature "the

I. C. Crăciun, *Positive Aging and Precarity*, International Perspectives on Aging 21, https://doi.org/10.1007/978-3-030-14255-1_10

precariatised mind" (Standing 2011). Nevertheless, the "precariatised mind" has not been clearly and thoroughly defined in previous research. Only some characteristics were described such as short-term thinking (i.e., the incapacity to think long-term), the incapacity to concentrate (Standing 2011), and a pessimistic outlook on life in general and low self-esteem (Malenfant et al. 2007). Even if enlightening, these features do not really explain what happens in the "precariatised mind" and especially not so in the context of aging. Standing (2011) in his book "*The precariat: the new dangerous class*" suggests that several behavioral, emotional and mental changes are happening with the spread of "precarization" (Standing 2011, p. 19). Findings from the present research on middle-aged precarious Germans add to the understanding of what goes on in the minds of precarious workers, what changes may be taking place and how they cope with precarity and aging. Several thought and behavior patterns came up frequently in the interviews with the middle-aged precarious Germans (see Chaps. 5, 6 and 7). Some of these such as short-term thinking patterns, lack of planning and the feeling of "no future" lend support to previous literature (Fonti 2017; Standing 2011). Nevertheless, the novelty of the present findings lies in the fact that they show how middle-aged individuals can turn some of these "shortcomings" of the precarious mind to their advantage in order to help them cope with aging in precarious times.

For example, short-term thinking and lack of planning are a consequence of the flexibility imposed by precarity (e.g., constantly being ready to change jobs, adapt to novel work schedules and work conditions, learn new skills etc.). However, precarious middle-aged Germans reconstructed flexibility as a necessary quality for survival in uncertain times. In their view, one needs to constantly "reinvent oneself" to adapt to the changing, insecure circumstances. The imposed flexibility was not simply regarded as an obstacle. Flexibility was used both as a coping mechanism (e.g., learn to be flexible and survive precarity) and as a positive thinking strategy by middle-aged precarious Germans. Having short-term work contracts, part-time jobs or being part of the working poor (Kalleberg 2013) could make people internalize flexibility and reframe it as an asset in their preparation for positive aging. Nevertheless, who fails to do so, may experience negative emotions (e.g., anxiety, depression, frustration) when dealing with precarity during the aging process (e.g., uncertain work and pension prospects, vulnerability of the aging body etc.).

Even if some middle-aged precarious individuals do experience a feeling of "*no future*" (Fonti 2017), and have no concrete expectations regarding their old age, the aging precariat is active in what concerns building up resources for old age. The interviewed precarious middle-aged Germans did not give in to depression or suicidal ideation as previous literature has found for other precarious work contexts (Kim et al. 2012; Min et al. 2015; Han et al. 2017). On the contrary, they consciously used positive reframing to motivate themselves to be active both in what concerns work and social relations. Furthermore, the fact that they were deliberately not planning was sometimes applied as a stress management strategy. Positive reframing was considered a more useful strategy in precarious circumstances as compared to planning for an uncertain future. Furthermore, the fact that precarious workers do not make long-term plans does not mean that they do not engage in short-term planning.

Precarious individuals have goals in life, but these are mostly centered on the present. According to existing research, the perception of a diminishing time perspective makes people concentrate on the most essential goals in their lives (Freund 2008). People usually tend to focus on the present and on goals that bring immediate pay-off when they have a limited time perspective (Freund 2008; Carstensen et al. 1999, 2003). Usually such short-term goals are emotionally meaningful, and people are motivated to shift their attention toward these when they think they have little time left to achieve their long-term goals (Carstensen et al. 2003). Middle-aged precarious Germans experience a limited time perspective for example due to their temporary work contract (e.g., six months contracts or one-year contracts etc.). Thus, they preferred to focus on emotional regulation by employing strategies such as cognitive reframing and making short-term plans instead of planning an uncertain future.

Because they are aware of the fragility of life (i.e., precarity), precarious middle-aged people focus on developing and maintaining social relations that have high emotional value such as those with family and close friends. However, they also keep the expansion of the social network as an active goal since they need new contacts and information for their job hunt. Precarious persons pay respect to meaningful social relations, lending support to the emotional selectivity theory (Carstensen et al. 1999, 2003) that assumes that when time perspectives are limited, people concentrate on their existing social relations and not necessarily on meeting new people. Nonetheless, present findings show that precarious individuals focus both on emotionally relevant relations and on expanding their social networks. Making new social contacts is important for finding work or improving job related skills, while the meaningful social relations are relevant for emotional support and even instrumental support (e.g., living in a shared flat).

Despite the expectation that precarious individuals would be passive in terms of old age preparations, middle-aged precarious Germans adapted their preparatory actions to fit the precarious circumstances. They engaged in activities concerning finding work, developing skills for being an attractive candidate on the job market in late life. Moreover, they actively focused on maintaining and expanding their social work-related networks. Such action patterns are connected to their belief that management of resources at present is more relevant than planning an uncertain future. Developing resources for old age was seen as a dynamic, long-term process, made up of pursuing a series of short-term goals. For example, in order to make sure they will have social support in old age, they developed their social networks at present. Potential social resources like family and friends are cultivated because they are important both at present and for the future. Social relations and networks represent social capital that could at some point compensate for a lack of financial capital in terms of preparation for a good old age. According to Phillipson (2013) it is important to distinguish between "bridging" and "bonding" social capital (Putnam 2000) and examine whether individuals form relations only with people who are similar to them (i.e., bonding) or also with people who are different (i.e., bridging). Findings show that precarious individuals tend to bond with other precarious people from the same age group. While this helps by building solidarity and providing social support (e.g., finding understanding for their problems), it may not be as effective as developing bridging social capital (Phillipson 2013; Putnam 2000). The latter would help precarious individuals to gain access to novel resources and maybe even form non-precarious identities.

According to the existing evidence base, one way of achieving a positive old age is by having positive views on aging, since these are associated with health in old age (Levy et al. 2002; Sargent-Cox et al. 2012; Wurm et al. 2007, 2010) as well as longevity (Kotter-Grühn et al. 2009). Precarious middle-aged Germans hold several negative views on aging (see for example Chaps. 5, 6, 7). Nevertheless, as discussed in previous chapters, precarious middle-aged Germans who took part in the study, developed their own versions of positive aging views. These novel views on aging were adapted to the precarious reality they lived in and comprised short-term planning to acquire resources (e.g., social networks, skills) and develop wisdom, self-acceptance and a general positive outlook on life. Precarious individuals internalized the idea that they want to be active, independent and healthy in old age. However, the way they define activity and independence is tailored to their precarious living and working conditions.

Another way of reaching a happy old age is by preparing it thoroughly, as recommended by active aging policy (Lassen and Moreira 2014; Walker 2009). As examined in Chaps. 7 and 9, precarious middle-aged Germans do prepare for old age in their own way that does not correspond entirely with active aging policy recommendations. For them, preparatory efforts include building social networks, developing family ties and romantic partnerships, strengthening the capacity to think in a positive way about life. All these aspects help build relevant resources for a happy old age in precarious times. For example, social relations and connectedness may prove more useful in precarious contexts than being too independent and autonomous. Social networks and the support they provide can help when the future seems uncertain or a person feels vulnerable. In a context of precarity, social relations may provide the basic conditions for independence in old age. Accepting the inherent vulnerability of human existence can help people mobilize their resources for old age.

Successful aging was described in the literature as an effective management of resources (Freund 2008). In accordance with this definition (Freund 2008), precarious individuals shape their development by choosing and pursuing a set of goals. In precarious times, people must select and follow goals that help them manage everyday precarity. Thus, precarious individuals prepare for old age by building and managing their resources. According to the SOC perspective (Baltes and Baltes 1990), one could interpret that precarious people are "gains oriented" since they are determined to accumulate resources to maximize gains at present rather than prevent losses in the future. Acquiring and accumulating resources is essential for survival in precarious times. The conservation of resources becomes of high relevance with advancing age and precarious individuals work on maintaining their assets (e.g., keeping good relations with their families, keeping their skills for work intact, staying healthy). In view of this interpretation of the SOC model (Freund 2008, 2006), one can say that precarious middle-aged individuals prioritize their goals (selection), to maximize emotional gains (optimization) and to prevent social and financial losses (compensation) by taking their diminishing resources into consideration. In this context, resources are defined as means to achieving goals (Freund and Riediger 2001). Because of experiencing precarity, people tend to evaluate their resources as being sparse, especially their financial assets or their emotional coping resources.

According to Freund (2008), successful aging is a lifelong process and the resources accumulated in earlier years can be used later on whenever they are needed. Temporal sequencing strategies (selecting one goal and postponing others to solve goal conflict) are more difficult to apply when the time perspective is limited such as in old age (Lang and Carstensen 2002) or in precarious circumstances. Focusing on goals that address central and similar or inter-reliant life domains (e.g., work and health, work and relaxation) could represent a solution. This strategy leads to goal facilitation and stronger motivation to pursue goals as well as a higher probability to achieve them. For example, if financial and time resources are scarce and precarious people do not afford to take a holiday, they take some free time around work. As Alexandra (one of the precarious workers, see Chaps. 5–7) explained in her interview, she goes to a conference and takes a few extra days to relax at the conference location. This is how she manages to travel to a new place, work, and also relax for a few days. Another example is Thomas (precarious worker, see Chaps. 5–7) who uses his volunteer activities to give meaning to his life, keep social contacts and potentially also find work.

Nevertheless, contrary to the idea that older people focus on fewer goals, precarious people must diversify their aims. They cannot concentrate just on one or two major goals since they need to stay flexible to survive precarity. This means they need to have more goals in order to focus on different ones at different times. While financially secure individuals train their bodies and minds to avoid losses, precarious people target an accumulation of resources to maximize gains for as long as possible. In conformity with the construal level theory (Trope and Liberman 2010), temporally proximate goals include a representation of the means, they are "do" goals and more suitable for precarious individuals than long-term abstract goals for an uncertain (i.e., precarious) future.

Even if precarity (Portacolone et al. 2018; Grenier and Phillipson 2018) and the active aging framework (Walker 2009; Lassen and Moreira 2014) seem difficult to reconcile, some defining features of precarity can be integrated in the active aging framework. For example, *flexibility* and *activity* are recommended by active aging policy and constitute attributes of precarity as well. Middle-aged precarious individuals are socially engaged and very active in their daily lives. They adapt to precarity by being flexible and focusing on their present goals. Their investment in resources for old age can be regarded as a form of proactive coping (Ouwehand et al. 2007) and thus one possible formula for successful aging preparation. Furthermore, because precarious individuals create their own strategies for preparing old age, one can say that at least the criteria of personalized preparation are met. Precarious workers do take responsibility for their old age and actively prepare by choosing specific means that are available to them. Nonetheless, inequalities between precarious and financially secure individuals may still arise since the first have fewer means and more obstacles for aging well.

All things considered, precarity helps to build peoples' strengths not just contributes to exposing their vulnerabilities. Individual actions develop at the intersection between social contexts and individuals' selves. Thus, what people make out of their precarious social context depends mostly on themselves and this is in line with the principle of taking responsibility for ones' own positive old age. However, at the same time, too much responsibility toward one's own old age may become stressful when living and working in precarious circumstances. Even if precarity does represent a new culture of risk that affects old age (Grenier et al. 2017; Grenier and Phillipson 2018), it also provides the opportunity to develop new skills that may prove useful for aging in precarious times. Nevertheless, in order to prevent the development of inequalities in old age, precarious individuals need help and guidance. In the following, I will explore some possible solutions of what can be done at individual and community level to foster positive aging for all.

Implications for Psychological Practice

Changes at Individual Level in Psychological Counselling

Both precarious and financially secure individuals need to find a meaning for their old age. However, precarious individuals face a double challenge because they need to invest both aging and the experience of precarity with meaning in order to experience well-being. They have to come to terms with being aging individuals living in precarious times (e.g., retreat of the welfare state, work and pension insecurity etc.). In contrast, for financially secure individuals finding meaning for old age is perhaps an easier task. For people who have sufficient material assets, retirement constitutes a somewhat "stable" condition associated with aging and provides a framework for organizing ones' life. Moreover, for financially secure individuals it is easier to internalize the positive aging discourse and use it to invest old age with meaning (see Chaps. 6 and 7).

Finding meaning within the experience of precarity is problematic since it concerns developing explanations for uncertainty and risk. Furthermore, the precarity of the aging experience is made worse by working and living in precarious conditions, resulting in the twofold challenge for precarious middle-aged individuals. However, even if precarious individuals need to cope with aging and precarity at the same time, they do not have to figure out everything by themselves. One relevant task for psychologists would be to help middle-aged precarious individuals to come up with solutions for coping with both precarity and aging. In the following, I will formulate some recommendations for psychological practice based on the findings from the research with precarious middle-aged Germans. These recommendations include identifying some of the potential features of the "precriatised mind" (Standing 2011) and devise solutions to help people cope with precarity and aging. Even if these solutions are based on the findings from the study conducted in Germany, this does not mean that they cannot be applied to other social and cultural contexts that are affected by precarity. Since precarity is a global phenomenon, similar problems and solutions will be found in different countries or social contexts affected by precarity.

Avoiding "mind traps" such as thinking there is "*no future,*" or one is "*too old to do something*" or that "*one has started too late to prepare for old age*" can help boost the motivation and prevent negative emotions in precarious middle-aged people.

One negative thought pattern that often came up during the interviews is that precarious individuals define themselves as a "*surprised generation.*" This means that precarious middle-aged Germans had no time to adjust to the changes in the work and retirement systems and had not accumulated financial resources to secure their old age. Blaming oneself for failing to prepare can be very harmful to preparatory actions because people would feel discouraged from preparing for an old age that seems unrealistic. Or as **Michael** (freelancer with no secure pension, see Chaps. 5–7) explained in relation to precarious work, if you are self-employed you have no boss to blame when things do not go well. When self-employed, a person needs to take responsibility for both success and failure at work. In time this can become very stressful.

Feelings of guilt concerning their current situation (i.e., working in a precarious job) were found in previous studies with precarious workers and unemployed individuals (Malenfant et al. 2007). Also, low self-esteem was associated with precarious work in several studies (Malenfant et al. 2007; Lewchuk et al. 2008; Kim et al. 2012). As discussed in previous chapters, positive images of aging inspired by the active aging policy can have a negative effect on preparation efforts. Namely, since precarious individuals cannot identify with the positive images of happy and active retirees, they may feel discouraged from planning their old age. Psychological counseling should aim to help precarious individuals avoid "feeling bad" about themselves if they do not "fit the profile" of the happy ager or if they have not accumulated enough material assets for old age.

As present findings have shown, precarious individuals place a lot of emphasis on their past choices, considering that their desire for freedom and independence has led them to choose precarious jobs instead of more secure, permanent jobs. Because they take responsibility for their life choices, precarious middle-aged individuals are "at risk" of blaming themselves for their decisions. However, the aging precariat did not know how the work and retirement situations would change and that they may end up without security for their old age. Negative emotions such as anger, depression, or anxiety can be avoided if precarious individuals are helped to reinterpret their past choices, emphasize the positive aspects of deciding for themselves (i.e., showing the capacity to decide independently, accepting their desire for freedom etc.), and avoid thinking about their work choices as being irrefutable mistakes.

Within the positive aging framework, forgiveness interventions are recommended when people must deal with losses or manage their grief (Hill 2011). For example, forgiveness interventions are used when people deal with difficult life transitions such as the pain caused by chronic diseases or the emotional burden of caregiving (Worthington 2006). In a similar way, forgiveness interventions may be applied with precarious individuals who need to "forgive" themselves for their past choices (e.g., taking on a precarious job, not finishing ones' studies, not putting enough effort into getting a permanent job, not having enough time for family or friends etc.) and prevent blaming themselves for becoming precarious. Instead of attributing their precarious situation to internal factors (e.g., lack of effort, absence of talent, lack of abilities etc.), they can accept that external precarious circumstances influence their lives. Forgiveness can help employ the necessary psychological resources to cope with the stress of precarity and preserve health and well-being. It could also benefit decision-making concerning aging relevant issues (e.g., health care, housing or pension options etc.). If people manage to forgive themselves for being precarious, they can make the best out of their assets for aging.

Use Flexibility as a Resource for Old Age

Flexibility represents a desired quality on the job market and in times of precarity (Lewchuk et al. 2008; Benach and Muntaner 2007) but also an obstacle for well-being since it can upset the work-life balance of precarious workers (Matilla-Santander et al. 2017; Süß and Sayah 2013). The way precarious middle-aged individuals cope with imposed work flexibility is that they "give up" planning long-term. This coping mechanism can be further trained within psychological counseling. People can develop flexible planning strategies. Instead of focusing on specific action and long-term plans, they can work on diversifying their goals and their readiness to change the means for goal achievement. For example, in terms of health behavior, coping planning (i.e., imagining obstacles for behavior adoption and thinking about solutions) may be more effective than forming action plans (i.e., thinking when, where and how to take up a certain behavior). Insisting that precarious individuals should plan and develop expectations for old age may not be the most effective strategy. It may prove more operative to encourage precarious people to continue to develop their assets for old age. Like this, they will be ready for "whatever comes" since this is how they mostly survive precarity in their daily lives. As I have discussed in previous chapters, precarious individuals construct flexibility, namely an ability to "constantly reinvent themselves," as an asset for surviving precarity. Instead of considering it a dangerous trait, flexibility could be further trained to help with selecting, optimizing, and compensating strategies for aging well.

According to positive psychology, flexibility refers to a person's' capacity to invoke novel behavioral and thinking strategies so as to foster better adaptation (Hill 2011). It involves balancing one's skills and personal resources to facilitate coping in stressful or demanding contexts (Hill 2011) such as precarity. Both flexibility in terms of thinking and behaving are components of the positive aging model proposed by Hill (2005, 2011). From a behavioral therapy perspective, flexibility alludes to the capacity to reframe automatic thoughts as a way to reduce their negative emotional impact.

Within the positive aging framework, gratitude represents a flexibility strategy that can help people to focus on positive attributes even when the events or circumstances are associated with loss. The underlying mechanism of gratitude interventions is similar to principles of cognitive behavioral therapy. One learns to reframe negative maladaptive automatic thoughts and disconnect these from negative emotions. In this process, latent emotional resources are generated, and well-being is preserved (Hill 2011). Studies on gratitude journaling interventions, showed that the effects included improved mood, a heightened perception of personal resources to engage in health behavior (Emmons and McCullough 2003). For precarious individuals, gratitude interventions may help concentrate on the positive aspects of both precarity and aging as well as help them generate the necessary resources for coping on a daily basis.

Become an Expert in Time Management

Time is a very important resource for precarious individuals since they have to organize different tasks simultaneously and maintain a feeling of being in control. As discussed in Chaps. 5, 6 and 8, precarious individuals lack the feeling that they are in control concerning their daily schedule or life in general. This feeling makes them avoid planning altogether. Nevertheless, they may benefit from learning some time management skills that can support them to organize their lives and achieve a work and life balance.

As examined in Chap. 5, sometimes the need for freedom and the desire to be one's own boss makes precarious individuals take on too much responsibility and end up not feeling in control of their time. The feeling of lacking control can lead to anxiety. However, better time management can help prevent anxiety and provide at least the illusion of control.

Time management implies solving the work-life balance that is more difficult to find for precarious individuals who do not have clear working schedules. Flexibility often means that there is no clear boundary between working hours and leisure time. Thus, precarious workers have to learn to set these limits themselves, without feeling guilty that they are not working or not actively looking for the next work opportunity. Time management in case of precarious workers is not just about meeting deadlines or being effective at work, but also about enjoying life in the process.

Work-life balance has become a frequently discussed topic in relation to flexible work-forms and contract workers (Süß and Sayah 2013). Precarious work poses many challenges in what concerns how different areas of a person's life (e.g., work, family, leisure etc.) can be coordinated to meet individual goals. The perception of an uncertain future, fluctuating payments, the requirement to be constantly available and the erosion of boundaries between work and private life are just some of the factors that can affect the well-being of precarious workers. Thus, finding ways to achieve work-life balance in precarious circumstances becomes a target for practice.

Change Negative Aging Stereotypes

As I have shown in previous chapters (see Chaps. 5, 6 and 7), precarious individuals hold a series of ***negative aging stereotypes that require changing***. The biggest threat in terms of negative views on aging would be not to think about old age at all. Thus, one needs to address the denial of aging, namely the postponement of thinking about old age or the internalization of agelessness. Precarity fosters particular fears (see Chap. 6) concerning aging that need to be tackled in psychological interventions. For example, the association of aging and illness or the denial of aging are thoughts that can be changed with the aid of psychological counseling. Addressing views on aging should not be restricted to interventions that promote physical activity. Several previous interventions attempted to change perceptions on aging to increase physical activity (Wolff et al. 2014; Klusmann et al. 2012; Sarkisian et al. 2007). While such interventions are much needed, they should be supplemented by other types of strategies and programs in order to prepare for a happy old age. Nevertheless, the techniques used by Wolff and colleagues (2014) to change negative views on aging may prove effective also in other settings. For example, Wolff and colleagues (2014) provided information on successful aging and included a component concerning views on aging in their physical activity intervention. Participants received information on positive aspects of aging, misconceptions about aging and the benefits of positive views on aging. Furthermore, participants were taught how to identify automatic unconscious negative thoughts related to aging and how to replace them with more positive views (Wolff et al. 2014). Such techniques can be applied also in interventions with precarious individuals to help them develop positive views on aging and cope in precarious settings. As discussed in Chaps 5, 6 and 7, precarity generates mainly negative views on aging. Nonetheless, people can be taught how to identify these negative automatic thoughts and replace them with more positive, adaptive thoughts concerning old age.

Changing views of aging might prove useful not just for health behavior but also for social engagement (Robertson and Kenny 2016). Other channels for changing negative views of aging have been proposed in the literature like changing the negative media representations or fostering intergenerational communication (Kotter-Grühn 2015). Such interventions are based on the assumption that changing societal views on aging will eventually lead to modifying also personal views of old age or self-perceptions of aging.

Address the Threat to the Self-identity

Precarious individuals have difficulties defining themselves by way of a certain job or career, through specific hobbies or retirement plans. For example, precarious middle-aged Germans defined themselves via their "constant change," their "rebellious" nature (as depicted in Chap. 5). Investing old age with meaning represents a challenge that all middle-aged adults face at some point, no matter if they are financially secure or work in precarious jobs. The negative aging stereotypes are part of the social reality that is shared by precarious and financially secure individuals alike (see Chaps. 5, 6 and 7). Even if financially secure individuals assume the positive aging model to a greater extent, this does not protect them entirely from lack of meaning regarding the aging processes or their work. For precarious persons who totally identify with the job, the threat to their self-identity in old age is quite large. Identifying with a precarious job, means that the person will identify with the very essence of precarity. Thus, interventions should aid precarious people to find other resources for building their self-identity (e.g., their personal qualities and strengths, hobbies etc.) outside the field of work.

Findings from one study have shown that interventions that help view retirement in a positive way and prepare for retirement, were associated with increased longevity (Lakra and Levy 2012). Lakra and colleagues (2012) found that financial preparation is not sufficient and that psychological preparation for retirement is also extremely relevant. Nevertheless, for individuals who have no retirement prospects such as many precarious people, the act of psychologically preparing for old age may still be highly beneficial. For instance, an intervention with older women from a poor neighborhood in Lisbon pointed out that it was relevant for such women to imagine a future and to plan and implement activities outside their daily routines (Paolletti and Gomes 2014). The workshop with the older women with low socioeconomic status targeted the discrepancy between what the women thought of themselves and their future and what they could actually make of their future and who they were as individuals. During the workshop, the women became aware of their thoughts about aging. After the identification of their views on aging, women were confronted with what they could actually still do, despite their old age (Paoletti and Gomes 2014). When asked about the problems encountered in implementing the workshop, the psychologist said that participants could express desires and dreams related to old age, but had difficulties making plans and translating wishes into action (Paoletti and Gomes 2014). This could be a situation that precarious individuals face. When thinking about aging, middle-aged precarious people may have wishes and dreams, but because of their precarious circumstances they might not see how they could make these wishes come true. Thus, a potential intervention aim would be to work with them toward finding solutions concerning how to transform their wishes into realizable plans.

Foster Positive Thinking

Positive reframing or optimistic thinking represent strengths that precarious individuals possess and use as coping mechanisms in precarious times. Keeping an optimistic viewpoint is considered part of the successful aging model (Hill 2005, 2011). It also represents a component of the positive aging views definition since it indicates that people imagine a positive future for themselves. Positive thinking implies having positive expectations concerning the future as well as a constructive present. It also implies focusing on the positive aspects even when being aware of the downside of things. Similar to persons who for one reason or another have lost resources (Hill 2011), precarious individuals can alter their definitions for successful aging and construct a positive aging representation that is realistic and attainable.

Both optimism and positive affect play a relevant role for healthy aging (Chida and Steptoe 2008; Lyyra et al. 2006; Kubzansky and Thurston 2007). Positive affect is associated with better cognitive functioning for memory and attention (Isen 2008; Yang 2013). A "positive bias" has been proven in case of older adults, in what concerns attention and memory tasks (Reed et al. 2014). Aging individuals, especially precarious workers might benefit from positive thinking training by using this positive bias as a starting point. This type of trainings can help people reduce or prevent negative emotions such as anxiety or depression that are usually associated with precarity. For example, cognitive bias modification trainings were shown to help with treatment of mood disorders (Woud and Becker 2014), anxiety treatment (MacLeod and Mathews 2012), or increase positive affect in young healthy participants (Holmes et al. 2008). In imagery cognitive bias modification, participants are taught to automatically imagine positive resolutions of ambiguous situations (Holmes et al. 2006, 2008). Participants are asked to repeatedly visualize positive images in response to ambiguous scenarios or images. In older adults this kind of training is said to be associated with increased optimism and with positively imaging the future (Murphy et al. 2015). Murphy and colleagues (2015) demonstrated that an imagery-based cognitive bias modification training helps reduce negative affect and increase optimism about the future in older adults. Precarious middle-aged individuals may also benefit from such cognitive imagery trainings in which they would be taught to visualize a positive old age. Another possible benefit would be that they would envisage positive resolutions to ambiguous situations such as precarious life situations. This may help to prevent precarious middle-aged people from feeling anxious or depressed in insecure work and life circumstances.

Develop Social Networks for Strengthening Social Support

Social networks are very important to precarious individuals and as discussed in chapters seven and nine, precarious workers look for ways to develop their social relations. Social networks help to prevent loneliness in old age. Precarity as we have seen in previous chapters tends to be related to loneliness and thus it becomes even more relevant for precarious individuals to prevent it. Thus, building social networks and maintaining existing social relations can help with having the necessary social support both at present and in the future. Some psychological strategies can foster the development of new social relations such as altruism interventions (Hill 2011). According to Hill (2011) altruism constitutes the background for volunteering activities. Studies have found that volunteering has multiple health benefits and increases longevity and well-being (Morrow-Howell et al. 2003). Thus, volunteering can help derive meaning and purpose in life and can be effective even when one's own health is declining (Shmotkin et al. 2003; Tan et al. 2006). As discussed in chapter four, some precarious individuals find meaning in volunteering and derive a feeling of well-being despite their precarious circumstances. Even if precarious people sometimes do not have time to volunteer outside their social networks or families, they still engage in actions for others and build social capital at the same time. It also provides an opportunity to exercise their altruism or to realize their generativity potential in old age (Schoklitsch and Baumann 2012).

All things considered, for encouraging social relations as well as for other intervention aims such as changing negative stereotypes about aging, policy interventions are needed to complement the effects of individual psychological counseling.

Implications for Active Aging Policy

Despite the globalization of the active aging policy idea, the question still remains open whether positive aging is "for all" or just for some privileged groups of individuals who can "afford" to age well. This question lies at the core of the positive aging policy criticism (Lassen and Moreira 2014; Carney and Gray 2015; Boudiny 2012).

In the German context there is an increased preoccupation with promoting positive aging (Mahne et al. 2017; Tesch-Römer and Wurm 2012) and several policy solutions have been designed. Among them, the promotion of positive and realistic images of aging has a high priority (Wurm and Huxhold 2012). Nevertheless, in order to be realistic, these novel images of aging need to take into account the new risks concerning old age (Grenier et al. 2017).

In the following, I discuss what solutions are created in terms of work and health policies, focusing on views on aging. Furthermore, I examine how mass-media could help with redesigning the views on aging as a consequence of policy or as a facilitator of policy recommendations.

Manage Age Discrimination on the Job Market

Planning for retirement represents a complex process that occurs over the whole life span of the individuals' working life and is influenced by a multitude of factors such as social norms, the workers' health, economic situations, institutional practices, or aging expectations. Retirement plans are extended decision-making processes that occur over the work life span (Szinovacz et al. 2013). In case of precarious workers, they reflect their work trajectories and the social context. Since retirement plans mirror the acquired assets (Szinovacz et al. 2013), precarious workers are at a disadvantage as compared to financially secure individuals. People with defined pension plans and high income can afford to retire earlier and lead a financially stress-free old age. However, this is not the case for precarious workers.

Work and retirement policies have changed in the last years to favor a higher retirement age and increase of labor force participation in old age (OECD 2011; Cahill et al. 2015; Berger 2009). Such changes are described as supply driven. As a result, there are more older workers seeking employment until and beyond the normal retirement age (Dingemans et al. 2015). However, most organizations are reluctant to retain or recruit older workers and younger workers are preferred over older workers when it comes to promotions or evaluation (Bal et al. 2011). In addition, there are individuals working in precarious jobs for whom the concept of retirement does not really exist. Precarious people are constantly looking for work, before and after the normal retirement age. While hunting for jobs, precarious individuals enter a competition for work with financially secure older people as well as younger generations of precarious workers. Moreover, it was shown that once unemployed, older people take longer to find a job since they face discrimination in their job search (Walker 2002). Postretirement work is only available to certain subgroups, and for example those who have experienced involuntary career exits have difficulties finding work past retirement age (Dingemans et al. 2015). There are still limited recruitment and retainment policies for older workers and decisions are made in very arbitrary ways (Oude Mulders et al. 2016; Oude Mulders et al. 2014). One study showed that employers used exit strategies much more often than they used developmental measures (e.g., training and promotion) for aging employees (Van Dalen et al. 2015). However, just encouraging late retirement (Walker 2006) is not enough, one needs to increase the eligibility and attractivity of older workers on the job market and combat discrimination of older people in work contexts (Boudiny 2012).

Top managers' norms about retirement age affect the organizations' recruitment and retention practices in what concerns older workers. The norms regarding age equality affect recruitment practices before the normal retirement age. Top managers' norms concerning retirement age affect retainment practices after the normal retirement age or the recruitment of retirees (Oude Mulders et al. 2016). The "normal" working age is defined as the interval between 20 to 65 (Oude Mulders et al. 2016). Norms are a relevant driver of work behavior and thus require changing if one wants to change retraining and recruitment practices for older workers. Despite the novel policies encouraging later retirement, work after the age of 65 is quite uncommon in Europe (Oude Mulders et al. 2016). Thus, managers have an important say regarding who gets hired and who can stay and work after the standard retirement age. In this context, changing the negative age stereotypes of managers becomes relevant. Precarious individuals may suffer from a double discrimination when searching for jobs, or even a triple one in case of precarious women. Organizations need to adapt to a diverse aging workforce and develop strategies to recruit different types of older employees. Nevertheless, changes in the retirement and social security systems are needed to ensure living conditions for those older workers who are not healthy enough to continue work or cannot find jobs. In terms of hiring policy, Boudiny (2012) suggests considering competence-based evaluations for older adults as compared to age-based criteria that can lead to discriminating older job candidates.

Intergenerational contact situations provide good opportunities for changing views on aging or old people. However, for precarious individuals, one common situation where they are in contact with younger individuals is when they compete for the same jobs. Thus, a clear "job market" needs to be defined for older people, especially for those living and working in precarious circumstances.

The feeling of "no future" (Fonti 2017) is sadly not an illusion until better work opportunities are created for older individuals or their preparatory efforts for old age facilitated. All in all, due to demographic changes and the retreat of the welfare state, the challenge for the 21st century is for organizations to find innovative strategies to retrain, value and invest in older workers (Moen et al. 2016). Although having an active older workforce is on the policy agenda of the EU (Loretto and Vickerstaff 2015; Walker 2002, 2006), better strategies are needed to integrate older workers and prevent age discrimination at work.

Health Policy for All- a Dream or a Possibility?

Health in old age represents a complex phenomenon and is influenced by several personal and social factors. Precarious workers face multiple health disadvantages that can lead to health inequalities in old age. First, precarious work (i.e., without benefits, with a set termination date, linked with a feeling of powerlessness) is associated with poor health (Ferrie et al. 2005; Benach and Muntaner 2007). Second, precarious individuals hold negative images of aging (Craciun and Flick 2014) which determine worse health (Levy 2009; Wurm et al. 2013). Policy interventions can address precarious work conditions as well as negative images of aging in order to promote health for precarious individuals.

In a previous section, I have discussed how images of aging can be used to influence individual health behavior. However, such individual behavioral and cognitive counseling initiatives are not enough to insure the health of precarious workers. Interventions that transform the work and living environment as well as changes in the healthcare policies are required. Initiatives such as age friendly cities (Warth 2015) provide a good base for creating an environment that provides equal health opportunities. It is important to create public spaces where people feel secure and where they are encouraged to adopt health behavior (e.g., parks, bicycle lanes etc.). Such public spaces should also stimulate community interaction since building new social networks is especially good for precarious individuals. For example, the availability of shared work spaces for freelances where they do not have to pay a fee or just a small symbolic contribution could help foster social relations in this group. Cultural activities funded by the state or spaces such as libraries, where people can come together and exchange ideas, also on an intergenerational level (intergenerational share ideas spaces) can constitute opportunities for precarious workers. Idea share centers could be founded for the wellness of the whole community and an intervention against precarity.

Policies concerning the price of healthy food are also relevant, since precarious people have fewer financial assets to invest in their healthy body. Health insurance is sometimes very expensive and when you are on an intermittent budget it is difficult to pay regularly. Nevertheless, health is a very significant asset for precarious people and they should benefit from this right all the time not just sometimes. Therefore, health insurance policies should be adapted to the needs of precarious individuals.

Changing negative aging stereotypes in individuals is not enough. Health professionals also need to change their negative aging stereotypes that can affect their communication with older patients (Walter et al. 2006; Craciun 2016, 2018; Craciun and Flick 2015). Health personnel usually have a lot of contact with older individuals, even if not always time to talk and explain medical terms (e.g., diagnoses, treatments etc.). Therefore, doctors and other health staffs need to be helped to translate active aging principles into comprehensible language for older people and integrate them easily into their practice. They might also require help to change their own negative views of aging that could flow into their work with older patients. There is a need to train doctors and care personnel concerning the topics of aging and health as well as illness prevention (Craciun 2016, 2018; Craciun and Flick 2015).

Redesigning Old Age for Mass Media

Active aging policy in Germany recognizes that images of aging are connected to the social context (Wurm and Huxhold 2012). It is also stipulated that aging policy should not promote unrealistic images of aging (Tesch-Römer and Wurm 2012), in line with the critics of the successful aging model (Boudiny 2012; Liang and Luo 2012; Minkler and Holstein 2008).

As discussed in previous chapters, in contexts of precarity all aging individuals risk having negative views on aging. According to stereotype embodiment theory (Levy 2009), people unconsciously integrate such negative aging stereotypes from their environment (in this case precarious work and life) that afterward affect their health. Thus, negative aging images influence bad health and the latter reinforce negative images of aging where old age is associated with illness (Wurm et al. 2013).

The discrepancy between negative and positive images of aging starts at a broader social level, where both types of images are promoted through mass media. Previous literature has identified clearly defined stereotypes in relation of the depiction of older people. These were portrayed as "needy and helpless" (Clarke 2009), as "invisible" or missing persons" (Clarke 2014), as "greedy and lazy third agers" that profit from the welfare state (Gilleard and Higgs 2010) or as "heroic elders" (Carney and Gray 2015). The positive images are usually of wise old men (Clarke 2014) and of age-defying older women (Carney and Gray 2015). In a similar way, media representations of later life range between a "good old age" characterized by health, independence and financial security and an "unsuccessful old age" connected with illness, loneliness and disability (Rozanova et al. 2006; McHugh 2003).

Mass media portrayals of aging are very important, since as Rozanova (2010) explains, they shape the understanding of aging and transmit the meaning of aging across generations. Rozanova (2010) explored the media portrayals of successful aging and found that most reflected a neoliberal discourse of positive aging. These images encouraged individuals to take responsibility for unsuccessful aging, construct successful aging as an individual choice and as "staying engaged" (Rozanova 2010). Such portrayals of successful aging can be recognized in the positive views on aging that financially secure and precarious individuals expressed in the current study. This probably happens because productive aging ideas dominate the precarious aging context where one needs to be active and independent to deal with the uncertainties of life in old age.

The inconsistency of aging images is not generated by aging individuals, but it is probably just assimilated from the current social discourses. However, the negative views on aging most likely gain precedence over the positive ones because the negative discourses (e.g., old age as illness and dependency) are most concordant with a reality of precarity (e.g., a vulnerable body, an uncertain work place etc.). Simply changing the portrayal of older adults in the media with more idealistic images only helps to increase the discrepancy between the promoted positive ideal and the precarious reality. Consequently, precarious individuals would feel they cannot ever reach the goals of positive aging and thus their self-efficacy toward goal achievement will decrease. Thus, just promoting positive views on aging without putting them in the context of precarity, can have the opposite effect of increasing negative views on old age in precarious contexts.

When designing policy interventions, one needs to take into account the social context. For example, one could promote realistic positive aging images or invest in resources to make positive aging available to all citizens (including those who live in precarity). Precarious individuals expect realistic portrayals of aging. They need to identify with the successful agers; thus, they need to be exposed to role models who are precarious but age in a positive way. This is not an impossible task. According to findings from the present qualitative research, middle-aged precarious people do find ways to age well and can serve as inspiration for similar individuals.

Among the things that need to be changed are the fixation on the aging body and the promotion of the denial of aging. Concerning the former, more emphasis could be placed on the aging mind and the benefits of experience on well-being in old age. This should also be done by taking gender into account (e.g., portray wise women not just wise men). Furthermore, the split between body and mind in what concerns aging has been criticized in the literature for being a form of ageism (Andrews 1999). Regarding the latter, the benefits of accepting old age as a normal stage of life can be thoroughly promoted. Many current images of aging encourage older people to look younger or not act their age. These ideas have been criticized for indicating a concealed form of ageism. It is suggested that they are indirectly implying old age is bad and ugly, while young age is good and beautiful (Smirnova 2012). McHugh

(2003) has called the coexistence of negative and anti-aging campaigns (e.g., selling creams to prevent wrinkles), bipolar ageism. According to McHugh (2003), the concept of agelessness hides the denial of disability and death and old age itself.

To summarize, even though it is important that aging individuals should stay fit and healthy, it is relevant that they learn to accept aging with all its advantages and disadvantages. Overall, showing several different role models that can reflect the heterogeneity of the aging experience and provide a wide range of possibilities to choose from when deciding how to prepare for a happy old age or invest old age with meaning.

Conclusions: The Way Forward

Two main questions have guided the arguments and theses of the present work. One refers to how precarity may shape the aging experience (i.e., views on aging and preparation for old age), while the other questions whether positive aging can exist in times of precarity. Regarding the first question, findings point out that precarity does shape views on aging at both an individual and social level. At an individual level, precarity generates a series of negative views on aging that represent an obstacle for a happy old age. However, in order to deal with the discrepancies between the social reality of aging and successful aging ideals, precarious workers develop alternative positive aging views as well as strategies to prepare for old age. Nevertheless, the case studies that were discussed and compared throughout this book show that there is a high risk of health and social inequalities emerging among aging individuals. Precarious workers face more obstacles when trying to achieve the active aging ideals (e.g., health, autonomy, social integration etc.).

On a social level, views on aging are shaped by precarity in the sense that older precarious individuals may face the double discrimination of being seen as old and precarious. Thus, they face the risk of marginalization on the basis of their age and their precarious circumstances. As previous literature has shown, baby boomers are perceived as a "threat" (van den Heuvel 2012) and the precariat (including the aging precariat) as a "dangerous" class (Standing 2011).

In what concerns whether positive aging can exist in precarious times, the present findings suggest that active aging ideals contradict the social reality of aging. However, alternative ways to age well are created and thus one may say precarity generates alternative positive aging goals and strategies to achieve these. Since the struggle to age well may come at a high cost for precarious individuals, interventions at individual and policy level should aim to help them achieve positive aging ideals and prevent inequalities.

The present book does not claim to answer all questions or recommend solutions to all problems related to precarity and aging. Without the ambitions of developing a new theory of precarious aging, it hopes to point out the relevance of asking novel questions regarding the links between positive aging and precarity. In addition to highlighting the importance of exploring the phenomenon of aging in precarious times, it attempts to offer a fresh psychological perspective on these issues. Furthermore, the book aims to show that in order to change the problems related to precarious aging, one needs to first change the frame of thinking about these issues (e.g., change views on aging). The thought patterns that generate difficulties related to aging in precarious circumstances require to be replaced with thought patterns (e.g., ideas, concepts, beliefs, theories etc.) that may generate solutions.

Since the book is based on qualitative research of mainly phycological topics, it does not offer complex answers for how to reform active aging policy. Several other studies address matters of policy change (Estes et al. 2003) or subjects such as precarity and aging from a sociological perspective (e.g., Phillipson 2013; Grenier et al. 2017; Millar 2017). Simple, three-step common sense psychological solutions about how to reach positive aging are also not provided in the current book. Nevertheless, based on the findings of the research with precarious middle-aged Germans, some ideas are outlined concerning how people can cope with precarity and age well. In addition, the book contributes to the understanding of how precarity may affect the aging experience and what role do views on aging play in the process of aging in precarious times.

Views on aging depend on the social context where they emerge and were studied in relation to education level and socioeconomic status (Wurm and Huxhold 2012). Cultural differences concerning views on aging were researched between countries (Löckenhoff et al. 2009), Eastern versus Western culture (Fung 2013), as differences between ethnic groups (Weintritt 2012) or even in relation to inter-regional differences within a country (Wolff et al. 2018). Nevertheless, since precarity represents a new culture of risk (Grenier et al. 2017) and a feature of present-day aging societies (Fonti 2017; Grenier et al. 2017; Standing 2011; Bauman 2007; Butler 2004), additional studies are needed concerning views on aging in the context of precarity (Craciun and Flick 2014).

The phenomenon of aging and precarity or as one may call it "*precarious aging*" should be further explored with qualitative and quantitative methods. Good examples of sociological and ethnographical studies are the works of Grenier and colleagues who focus on precarity and aging (Grenier et al. 2017; Grenier and Phillipson 2018) or Portacolone who explores the experience of living alone in old age (Portacolone 2013; Portacolone et al. 2018). Future qualitative or quantitative studies can examine meanings of precarity for different age groups or investigate how the experiences of aging women differ from those of aging men in precarious contexts. Such studies can help refine the concept of precarious aging. Additionally, qualitative findings can inform quantitative research on the relation between precarity and health (i.e., physical and mental) and aging processes or even longevity within longitudinal design studies. Results from both quantitative and qualitative research would contribute to developing a model of precarious aging. Furthermore, even if positive aging is possible in precarious times, one question still remains open, namely: how can a precarious aging model be integrated in the successful aging paradigm?

The present book used several case studies of precarious and financially secure middle-aged Germans to explore relations between precarity, views on aging and positive aging. Therefore, no generalization claims are being made starting from these findings. Future research can look into precarity as a culture of risk that can influence the aging experience and views on aging on a global scale. Since precarity is establishing itself as a culture of risk around the globe, future studies will address the issue of precarity and aging to a larger extent than inquiries about intercultural differences concerning aging views.

Age itself can play a role in how individuals experience precarity. If having a precarious job in high school may not affect health and well-being since one can consider it "normal" work at that life stage, the situation changes for young or middle-aged adults. Authors such as Standing (2011) have distinguished between the rise of a young precariat as well as the existence of an older precariat. Future studies can compare the experience of precarity and its effects on health and well-being in young versus old precarious individuals. Such studies would help understand the experience of aging as well as how precarity is perceived at various life stages.

Links between precarity and aging need to be studied from a psychological perspective in addition to a sociological or cultural and political one. Psychologists need to understand how the "*precariatised mind*" (Standing 2011) works, how the stress of precarity affects aging individuals and how they can cope with it. The present work provides the first steps in this direction. On a theoretical level, new theories of positive aging are needed to match the requirements of aging in precarious times. Innovative successful aging models as well as research on views on aging will need to integrate the concept of precarity. On a practical level, we need action to improve people's lives and prevent social inequalities. One of the main challenges for future research is to figure out how positive aging can be designed in precarious times.

Better ways to define precarity on a psychological level are needed. Precarity as a characteristic of work life may prove reductionistic. Even though this information is used in the present research to form categories, it is difficult to delimitate the precariat as a class. As critiques of Standings' "class in the making" theory (2011) have stated, it is difficult to describe the precariat as a homogenous group (Millar 2017). For the present book, most illustrative cases were of precarious individuals with higher education rather than low-paid workers performing menial jobs. However, this was not intentional, it rather happened that for this type of qualitative research more educated individuals declared themselves willing to be interviewed. Starting from the present findings, it is difficult to conclude that the aging precariat is characterized by anger, anomie, anxiety, and alienation (Standing 2011). Although some of these emotions do characterize precarious individuals, they are not always predominant, and people are trying to control these negative feelings and change their circumstances. Therefore, their effort should be acknowledged. Furthermore, from a psychological point of view, it is not advisable to label a group as "dangerous." Rather, it is better to describe the precarious circumstances as being "dangerous" and attempt to change these instead of encouraging fears concerning the group of precarious individuals (e.g., the aging precariat being a threat to young people or the functioning of the health or pension system).

Previous research has pointed out that precarity can be understood as a political notion (Millar 2017). Precarity possess a mobilizing capacity that has been made visible from the EuroMayDay protests in the early 2000s until contemporary demonstrations in 2018 such as the "Gilets Jaunes" movement in France and other countries. However, besides being a threat, precarity may also represent a resource in some circumstances. In addition to mobilizing individuals to protest and attempt to improve their lives, precarity also "moves people toward others" making them more empathic toward the suffering of others (Millar 2017; Butler 2004). Present findings have also pointed out an increase in empathy in what concerns the needs and problems of other people in precarious circumstances. Precarity makes people more aware of other's needs and increases their understanding of the problems inherent to dealing with both aging and life's uncertainty in general. Nevertheless, even if precarity may sometimes constitute a resource, one needs to fight against the spread and normalization of insecurity that encourages individuals to self-regulate their precarious circumstances (Millar 2017) or in this case aging.

All things considered, the concept of precarity helps shift the focus on the social context when studying aging (Grenier and Phillipson 2018). Starting from this premise, the present book argues that it is relevant to study the individual perceptions of precarity and aging and find out what happens in the "precariatised mind." This may help identify resources for aging in precarious times or even figure out how to turn precarity itself into a resource for positive aging.

References

Andrews, M. (1999). The seductiveness of agelessness. *Ageing & Society, 19*(3), 301–318. https://doi.org/10.1017/s0144686x99007369.

Bal, A. C., Reiss, A. E. B., Rudolph, C. W., & Baltes, B. B. (2011). Examining positive and negative perceptions of older workers: A meta-analysis. *The Journals of Gerontology Series B: Psychological Sciences and Social Sciences, 66B*(6), 687–698. https://doi.org/10.1093/geronb/gbr056.

Baltes, P. B., & Baltes, M. M. (Eds.). (1990). *Successful aging*. Cambridge University Press. https://doi.org/10.1017/cbo9780511665684.

Bauman, Z. (2007). *Liquid times: Living in an age of uncertainty*. Cambridge: Polity Press.

Benach, J., & Muntaner, C. (2007). Precarious employment and health: developing a research agenda. *Journal of Epidemiology and Community Health, 61*(4), 276–277. https://doi.org/10.1136/jech.2005.045237.

Berger, E. D. (2009). Managing age discrimination: An examination of the techniques used when seeking employment. *The Gerontologist, 49*(3), 317–332. https://doi.org/10.1093/geront/gnp031.

Boudiny, K. (2012). "Active ageing": from empty rhetoric to effective policy tool. *Ageing & Society, 33*(06), 1077–1098. https://doi.org/10.1017/s0144686x1200030x.

Butler, J. (2004). *Precarious life: The powers of mourning and violence*. London: Verso.

Cahill, K. E., Giandrea, M. D., & Quinn, J. F. (2015). Are gender differences emerging in the retirement patterns of the early boomers? *SSRN Electronic Journal*. https://doi.org/10.2139/ssrn.2645853.

Carney, G. M., & Gray, M. (2015). Unmasking the "elderly mystique": Why it is time to make the personal political in ageing research. *Journal of Aging Studies, 35,* 123–134. https://doi.org/10.1016/j.jaging.2015.08.007.

Carstensen, L. L., Fung, H. H., & Charles, S. T. (2003). *Motivation and Emotion, 27*(2), 103–123. https://doi.org/10.1023/a:10245698032304.

Carstensen, L. L., Isaacowitz, D. M., & Charles, S. T. (1999). Taking time seriously: A theory of socioemotional selectivity. *American Psychologist, 54,* 165–181.
Chida, Y., & Steptoe, A. (2008). Positive psychological well-being and mortality: A quantitative review of prospective observational studies. *Psychosomatic Medicine, 70*(7), 741–756.
Clarke, J. (2009). The case of the missing person: Alzheimer's disease in mass print magazines 1991–2001. *Health Communication, 19*(3), 269–276.
Clarke, L. H., Erica V. Bennett, C. L. (2014). Aging and masculinity: Portrayals in men's magazines. *Journal of Aging Studies*, 31 on pages 26 to 33.
Craciun, C., & Flick, U. (2014). "I will never be the granny with rosy cheeks": Perceptions of aging in precarious and financially secure middle-aged Germans. *Journal of Aging Studies, 29,* 78–87. https://doi.org/10.1016/j.jaging.2014.01.003.
Craciun, C., & Flick, U. (2015). "I want to be 100 years old, but I smoke too much": Exploring the gap between positive aging goals and reported preparatory actions in different social circumstances. *Journal of Aging Studies, 35,* 49–54. https://doi.org/10.1016/j.jaging.2015.07.006.
Craciun, C. (2016). "They Do Not Know How To Use Their Health Insurance Card": Implementation of active aging policy and aging perceptions in romanian general practitioners. *Ageing International, 41*(3), 254–264. https://doi.org/10.1007/s12126-015-9239-x.
Crăciun, I. C. (2018). Aging in precarious times. *Alter Und Geschlecht*, 163–175. https://doi.org/10.1007/978-3-658-20872-1_11.
Dingemans, E., Henkens, K., & van Solinge, H. (2015). Access to bridge employment: Who finds and who does not find work after retirement? *The Gerontologist, 56*(4), 630–640. https://doi.org/10.1093/geront/gnu182.
Emmons, R. A., & McCullough, M. E. (2003). Counting blessings versus burdens: An experimental investigation of gratitude and subjective well-being in daily life. *Journal of Personality and Social Psychology, 84*(2), 377–389. https://doi.org/10.1037/0022-3514.84.2.377.
Estes, C. L., Biggs, S., & Phillipson, C. (2003). *Social theory, social policy and ageing: A critical introduction.* London: Open University Press (US: McGraw-Hill).
Ferrie, J. E., Shipley, M. J., Newman, K., Stansfeld, S. A., & Marmot, M. (2005). Self-reported job insecurity and health in the Whitehall II study: potential explanations of the relationship. *Social Science and Medicine, 60*(7), 1593–1602. https://doi.org/10.1016/j.socscimed.2004.08.006.
Foster, L., & Walker, A. (2014). Active and successful aging: A European policy perspective. *The Gerontologist, 55*(1), 83–90. https://doi.org/10.1093/geront/gnu028.
Fonti, A. (2017). *General theory of the Precariat. Great Recession, Revolution, Reaction.* Downloaded from http://networkcultures.org/publications.
Freund, A. M., & Riediger, M. (2001). What I have and what I do: The role of resource loss and gain throughout life. *Applied Psychology: An International Review, 50,* 370–380.
Freund, A. M. (2006). Differential motivational consequences of goal focus in younger and older adults. *Psychology and Aging, 21,* 240–252.
Freund, A. M. (2008). Successful aging as management of resources: The role of selection, optimization, and compensation. *Research in Human Development, 5*(2), 94–106. https://doi.org/10.1080/15427600802034827.
Fung, H. H. (2013). Aging in culture. *The Gerontologist, 53*(3), 369—377. https://doi.org/10.1093/geront/gnt024.
Gilleard, C., & Higgs, P. (2010). Aging without agency: Theorizing the fourth age. *Aging and Mental Health, 14*(2), 121–128. http://dx.doi.org/10.1080/13607860903228762.
Grenier, A., & Phillipson, C. (2018). Precarious aging: Insecurity and risk in late life. *Hastings Center Report, 48,* S15–S18. https://doi.org/10.1002/hast.907.
Grenier, A., Phillipson, C., Laliberte Rudman, D., Hatzifilalithis, S., Kobayashi, K., & Marier, P. (2017). Precarity in late life: Understanding new forms of risk and insecurity. *Journal of Aging Studies, 43,* 9–14. https://doi.org/10.1016/j.jaging.2017.08.002.
Han, K.-M., Chang, J., Won, E., Lee, M.-S., & Ham, B.-J. (2017). Precarious employment associated with depressive symptoms and suicidal ideation in adult wage workers. *Journal of Affective Disorders, 218,* 201–209. https://doi.org/10.1016/j.jad.2017.04.049.

Hill, R. D. (2005). *Positive aging: A guide for mental health care professionals and consumers.* New York: W.W. Norton.

Hill, R. D. (2011). A positive aging framework for guiding geropsychology interventions. *Behavior Therapy, 42*(1), 66–77. https://doi.org/10.1016/j.beth.2010.04.006.

Holmes, E. A., Mathews, A., Mackintosh, B., & Dalgleish, T. (2008). The causal effect of mental imagery on emotion assessed using picture-word cues. *Emotion, 8*(3), 395–409. https://doi.org/10.1037/1528-3542.8.3.395.

Holmes, E. A., Mathews, A., Dalgleish, T., & Mackintosh, B. (2006). Positive interpretation training: Effects of mental imagery versus verbal training on positive mood. *Behavior Therapy, 37*(3), 237–247. https://doi.org/10.1016/j.beth.2006.02.002.

Isen, A. M. (2008). Some ways in which positive affect influences decision making and problem solving. In: M. Lewis, J. Haviland-Jones, L. F., & Barrett (Eds.), *Handbook of Emotions* (3rd ed.) NewYork: Guilford Press.

Kalleberg, A. L. (2013). Book review symposium: Response to reviews of Arne L Kalleberg, Good Jobs, Bad Jobs: The rise of polarized and precarious employment systems in the United States, 1970s to 2000s. *Work, Employment & Society, 27*(5), 896–898. https://doi.org/10.1177/0950017013480403.

Kim, I. H., Muntaner, C., Vahid Shahidi, F., Vives, A., Vanroelen, C., & Benach, J. (2012). Welfare states, flexible employment, and health: A critical review. *Health Policy, 104*(2), 99–127. https://doi.org/10.1016/j.healthpol.2011.11.002.

Klusmann, V., Evers, A., Schwarzer, R., & Heuser, I. (2012). Views on aging and emotional benefits of physical activity: Effects of an exercise intervention in older women. *Psychology of Sport and Exercise, 13,* 236–242. https://doi.org/10.1016/j.psychsport.2011.11.001.

Kotter-Grühn, D., Kleinspehn-Ammerlahn, A., Gerstorf, D., & Smith, J. (2009). Self-perceptions of aging predict mortality and change with approaching death: 16-year longitudinal results from the Berlin Aging Study. *Psychology and Aging, 24,* 654–667. https://doi.org/10.1037/a0016510.

Kotter-Grühn, D. (2015). Changing negative views of aging: implications for intervention and translational research. *Annual Review of Gerontology and Geriatrics, 35*(1), 167–186. https://doi.org/10.1891/0198-8794.35.167.

Kubzansky, L. D., & Thurston, R. C. (2007). Emotional vitality and incident coronary heart disease. *Archives of General Psychiatry, 64*(12), 1393. https://doi.org/10.1001/archpsyc.64.12.1393.

Lakra, D. C., NG, R., & Levy B. R. (2012). Increased longevity from viewing retirement positively. *Ageing and Society, 32*(08), 1418–1427. https://doi.org/10.1017/s0144686x11000985.

Lang, F. R., & Carstensen, L. L. (2002). Time counts: Future time perspective, goals, and social relationships. *Psychology and Aging, 17*(1), 125–139. https://doi.org/10.1037/0882-7974.17.1.125.

Lassen, A. J., & Moreira, T. (2014). Unmaking old age: Political and cognitive formats of active ageing. *Journal of Aging Studies, 30,* 33–46. https://doi.org/10.1016/j.jaging.2014.03.004.

Levy, B. R., Slade, M. D., Kunkel, S. R., & Kasl, S. V. (2002). Longevity increased by positive self-perceptions of aging. *Journal of Personality and Social Psychology, 83*(2), 261–270. https://doi.org/10.1037/0022-3514.83.2.261.

Levy, B. (2009). Stereotype Embodiment. *Current directions in psychological science, 18*(6), 332–336. https://doi.org/10.1111/j.1467-8721.2009.01662.x.

Lewchuk, W., Clarke, M., & de Wolff, A. (2008). Working without commitments: Precarious employment and health. *Work, Employment & Society, 22*(3), 387–406. https://doi.org/10.1177/0950017008093477.

Liang, J., & Luo, B. (2012). Toward a discourse shift in social gerontology: From successful aging to harmonious aging. *Journal of Aging Studies, 26*(3), 327–334.

Loretto, W., & Vickerstaff, S. (2015). Gender, age and flexible working in later life. *Work, Employment & Society, 29*(2), 233–249. https://doi.org/10.1177/0950017014545267.

Löckenhoff, C. E., De Fruyt, F., Terracciano, A., McCrae, R. R., De Bolle, M., Costa, P. T. Jr, …Yik, M. (2009). Perceptions of aging across 26 cultures and their culture-level associates. *Psychology and Aging, 24,* 941–954. https://doi.org/10.1037/a0016901.

Lyyra, T. M., Tormakangas, T. M., Read, S., Rantanen, T., & Berg, S. (2006). Satisfaction with present life predicts survival in octogenarians. *The Journals of Gerontology. Series b, Psychological Sciences and Social Sciences, 61*(6), P319–P326.

MacLeod, C., & Mathews, A. (2012). Cognitive bias modification approaches to anxiety. *Annual Review of Clinical Psychology, 8*(1), 189–217. https://doi.org/10.1146/annurev-clinpsy-032511-143052.

Mahne, K., Wolff, J. K., Simonson, J., & Tesch-Römer, C. (2017). Altern im Wandel: Zwei Jahrzehnte Deutscher Alterssurvey [Changes in Aging: Two Decades German Survey of Aging]. In K. Mahne et al. (Eds.) *Altern Im Wandel*, 11–28. https://doi.org/10.1007/978-3-12502-8_1.

Malenfant, R., LaRue, A., & Vézina, M. (2007). Intermittent work and well-being. *Current Sociology, 55*(6), 814–835. https://doi.org/10.1177/0011392107081987.

Matilla-Santander, N., Lidón-Moyano, C., González-Marrón, A., Bunch, K., Martín-Sánchez, J. C., & Martínez-Sánchez, J. M. (2017). Attitudes toward working conditions: are European Union workers satisfied with their working hours and work-life balance? *Gaceta Sanitaria.* https://doi.org/10.1016/j.gaceta.2017.10.006.

McHugh, K. E. (2003). Three faces of ageism: Society, image and place. *Ageing & Society, 23*(02), 165–185. https://doi.org/10.1017/s0144686x02001113.

Millar, K. M. (2017). Toward a critical politics of precarity. *Sociology Compass, 11*(6), e12483. https://doi.org/10.1111/soc4.12483.

Min, K.-B., Park, S.-G., Hwang, S. H., & Min, J.-Y. (2015). Precarious employment and the risk of suicidal ideation and suicide attempts. *Preventive Medicine, 71,* 72–76. https://doi.org/10.1016/j.ypmed.2014.12.017.

Minkler, M., & Holstein, M. B. (2008). From civil rights to civic engagement? Concerns of two older critical gerontologists about a "new social movement" and what it portends. *Journal of Aging Studies, 22*(2), 196–204. https://doi.org/10.1016/j.jaging.2007.12.003.

Morrow-Howell, N., Hinterlong, J., Rozario, P. A., & Tang, F. (2003). Effects of volunteering on the well-being of older adults. *The Journals of Gerontology: Series B, 58*(3), S137–S145. https://doi.org/10.1093/geronb/58.3.s137.

Moen, P., Kojola, E., Schaefers, K. (2016). Organizational change around an older workforce. *The Gerontologist, 57*(5), 847–856.

Murphy, S. E., Clare O'Donoghue, M., Drazich, E. H. S., Blackwell, S. E., Christina Nobre, A., & Holmes, E. A. (2015). Imagining a brighter future: The effect of positive imagery training on mood, prospective mental imagery and emotional bias in older adults. *Psychiatry Research, 230*(1), 36–43. https://doi.org/10.1016/j.psychres.2015.07.059.

Ouwehand, C., de Ridder, D. T. D., & Bensing, J. M. (2007). A review of successful aging models: Proposing proactive coping as an important additional strategy. *Clinical Psychology Review, 27*(8), 873–884. https://doi.org/10.1016/j.cpr.2006.11.003.

OECD. (2011). *Pensions at a glance 2011: Retirement-income systems in OECD and G20 countries.* OECD Publishing. https://doi.org/10.1787/pension_glance-2011-en.

Oude Mulders, J., van Dalen, H. P., Henkens, K., & Schippers, J. (2014). How likely are employers to rehire older workers after mandatory retirement? A vignette study among managers. *De Economist, 162*(4), 415–431. https://doi.org/10.1007/s10645-014-9234-8.

Oude Mulders, J., Henkens, K., & Schippers, J. (2016). European top managers' age-related workplace norms and their organizations' recruitment and retention practices regarding older workers. *The Gerontologist*, gnw076. https://doi.org/10.1093/geront/gnw076.

Paoletti, I., & Gomes, S. (2014). Future talk in later life. *Journal of Aging Studies, 29,* 131–141. https://doi.org/10.1016/j.jaging.2014.01.005.

Phillipson, C. (2013). *Ageing*. Polity Press, Cambridge, 2013, 218 pp., ISBN 13: 978 0 7456 3085 4.

Putnam, R. D. (2000). *Bowling alone*. New York: Simon & Schuster.

Portacolone, E. (2013). The notion of precariousness among older adults living alone in the U.S. *Journal of Aging Studies, 27*(2), 166–174. https://doi.org/10.1016/j.jaging.2013.01.001.

Portacolone, E., Rubinstein, R. L., Covinsky, K. E., Halpern, J., & Johnson, J. K. (2018). The Precarity of Older Adults Living Alone With Cognitive Impairment. *The Gerontologist*. https://doi.org/10.1093/geront/gnx193.

Reed, A. E., Chan, L., & Mikels, J. A. (2014). Meta-analysis of the age-related positivity effect: Age differences in preferences for positive over negative information. *Psychology and Aging, 29*(1), 1–15. https://doi.org/10.1037/a0035194.

Robertson, D. A. & Kenny R. A. (2016). "I'm too old for that" — The association between negative perceptions of aging and disengagement in later life. *Personality and Individual Differences, 100,* 114–119.

Rozanova, J., Northcott, H., & McDaniel, S. A. (2006). Seniors and portrayals of intra-generational and inter-generational inequality in The Globe & Mail. *Canadian Journal on Aging, 25*(4), 373–386.

Rozanova, J. (2010). Discourse of successful aging in The Globe & Mail: Insights from critical gerontology. *Journal of Aging Studies, 24*(4), 213–222. https://doi.org/10.1016/j.jaging.2010.05.001.

Sargent-Cox, K. A., Anstey, K. J., & Luszcz, M. A. (2012). Change in health and self-perceptions of aging over 16 years: The role of psychological resources. *Health Psychology, 31*(4), 423–432. https://doi.org/10.1037/a0027464.

Sarkisian, C. A., Prohaska, T. R., Davis, C., & Weiner, B. (2007). Pilot test of an attribution retraining intervention to raise walking levels in sedentary older adults. *Journal of the American Geriatrics Society, 55,* 1842–1846. https://doi.org/10.1111/j.1532-5415.2007.01427.x.

Shmotkin, D., Blumstein, T., & Modan, B. (2003). Beyond keeping active: Concomitant of being a volunteer. *Psychology and Aging, 18,* 602–607.

Schoklitsch, A., & Baumann, U. (2012). Generativity and aging: A promising future research topic? *Journal of Aging Studies, 26*(3), 262–272. https://doi.org/10.1016/j.jaging.2012.01.002.

Smirnova, M. H. (2012). A will to youth: The woman's anti-aging elixir. *Social Science and Medicine, 75*(7), 1236–1243. https://doi.org/10.1016/j.socscimed.2012.02.061.

Szinovacz, M. E., Martin, L., & Davey, A. (2013). Recession and expected retirement age: Another look at the evidence. *The Gerontologist, 54*(2), 245–257. https://doi.org/10.1093/geront/gnt010.

Standing, G. (2011). *The Precariat: The new dangerous class*. London: Bloomsbury Academic.

Süß, S., & Sayah, S. (2013). Balance between work and life: A qualitative study of German contract workers. *European Management Journal, 31*(3), 250–262. https://doi.org/10.1016/j.emj.2012.12.003.

Tan, E. J., Xue, Q.-L., Li, T., Carlson, M. C., & Fried, L. P. (2006). Volunteering: A physical activity intervention for older adults – The experience corp program in *Baltimore*. *Journal of Urban Health, 83,* 954–969.

Tesch-Römer, C., & Wurm, S. (2012). Research on active aging in Germany. *GeroPsych, 25*(3), 167–170. https://doi.org/10.1024/1662-9647/a000067.

Trope, Y., & Liberman, N. (2010). "Construal-level theory of psychological distance": Correction to Trope and Liberman (2010). *Psychological Review, 117*(3), 1024. https://doi.org/10.1037/a0020319.

Van Dalen, H. P., Henkens, K., & Wang, M. (2015). Recharging or retiring older workers? Uncovering the Age-based strategies of european employers. *The Gerontologist, 55*(5), 814–824. https://doi.org/10.1093/geront/gnu048.

Van den Heuvel, W. J. (2012). Discrimination against older people. *Reviews in Clinical Gerontology, 22*(04), 293–300. https://doi.org/10.1017/s095925981200010x.

Walter, U., Flick, U., Neuber, A., Fischer, C., & Schwartz, F. W. (2006). *Alt und Gesund? Altersbilder und Präventionskonzepte in der ärztlichen und pflegerischen Praxis*. [Old and healthy? Views on aging and prevention in health and care practice]. VS Verlag für Sozialwissenschaften.

Walker, A. (2009). Commentary: The emergence and application of active aging in Europe. *Journal of Aging & Social Policy, 21*(1), 75–93. https://doi.org/10.1080/08959420802529986.

Walker, A. (2006). Aging and politics: An international perspective. In R. Binstock & L. George (Eds.), *Handbook of aging and the social sciences* (pp. 339–359). Burlington, MA: Elsevier Academic Press.
Walker, A. (2002). A strategy for active ageing. *International Social Security Review, 55*(1), 121–140.
Warth, L. (2015). The WHO global network of age-friendly cities and communities: Origins, developments and challenges. *Age-Friendly Cities and Communities in International Comparison*, 37–46. https://doi.org/10.1007/978-3-319-24031-2_3.
Weintritt, O. (2012). Altersbilder im Islam und unter Muslimen in Deutschland und Körperbilder im Islam. [Views on aging in Islam and among Muslims in Germany and body images in Islam] In F. Berner et al. (Eds.) Individuelle Und Kulturelle Altersbilder [Individual and cultural views on aging], 231–287. https://doi.org/10.1007/978-3-93286-6_6.
Wolff, J. K., Warner, L. M., Ziegelmann, J. P., & Wurm, S. (2014). What do targeting positive views on ageing add to a physical activity intervention in older adults? Results from a randomised controlled trial. *Psychology & Health, 29,* 915–932. https://doi.org/10.1080/08870446.2014.896464.
Wolff, J. K., Beyer, A.-K., Wurm, S., Nowossadeck, S., & Wiest, M. (2018). Regional impact of population aging on changes in individual self-perceptions of aging: Findings from the German ageing survey. *The Gerontologist, 58,* 47–56. https://doi.org/10.1093/geront/gnx127.
Worthington, E. L., Jr. (2006). *Forgiveness and reconciliation: Theory and application*. New York: Routledge.
Woud, M. L., & Becker, E. S. (2014). Editorial for the special issue on cognitive bias modification techniques: An introduction to a time traveller's tale. *Cognitive Therapy and Research, 38*(2), 83–88. https://doi.org/10.1007/s10608-014-9605-0.
Wurm, S., Tesch-Römer, C., & Tomasik, M. J. (2007). Longitudinal findings on aging-related cognitions, control beliefs, and health in later life. *The Journals of Gerontology, Series B: Psychological Sciences and Social Sciences, 62,* 156–164. https://doi.org/10.1093/geronb/62.3.P156.
Wurm, S., Tomasik, M. J., & Tesch-Römer, C. (2010). On the importance of a positive view on ageing for physical exercise among middle-aged and older adults: Cross-sectional and longitudinal findings. *Psychology & Health, 25*(1), 25–42. https://doi.org/10.1080/08870440802311314.
Wurm, S., Warner, L. M., Ziegelmann, J. P., Wolff, J. K., & Schüz, B. (2013). How do negative self-perceptions of aging become a self-fulfilling prophecy? *Psychology and Aging, 28*(4), 1088–1097. https://doi.org/10.1037/a0032845.
Wurm, S., & Huxhold, O. (2012). Sozialer Wandel und individuelle Entwicklung von Altersbildern [Social change and individual devolvement of views on aging]. In F. Berner et al. (Eds.) *Individuelle und Kulturelle Altersbilder* [Individual and cultural views on aging], 27–69. https://doi.org/10.1007/978-3-531-93286-6_2.
WHO. (2002). *Active ageing: A policy framework*. Geneva: Author.
Yang, S. (2013). Wisdom and good lives: A process perspective. *New Ideas in Psychology, 31*(3), 194–201. https://doi.org/10.1016/j.newideapsych.2013.03.001.

Index

I. C. Crăciun, *Positive Aging and Precarity*, International Perspectives on Aging 21, https://doi.org/10.1007/978-3-030-14255-1

The manufacturer's authorised representative in the EU is Springer Nature Customer Service Centre GmbH, Europaplatz 3, 69115 Heidelberg, Germany. If you have any concerns regarding our products, please contact ProductSafety@springernature.com

Printed and bound by CPI Group (UK) Ltd, Croydon, CR0 4YY
30/09/2026
02225444-0005